ELEMENTARY TOPOLOGY

A Combinatorial and Algebraic Approach

ELEMENTARY TOPOLOGY
A Combinatorial and Algebraic Approach

Reprinted with corrections

DONALD W. BLACKETT

Department of Mathematics,
Boston University,
Massachusetts, USA.

1982

ACADEMIC PRESS
A Subsidiary of Harcourt Brace Jovanovich, Publishers
London · New York · Paris · San Diego · San Francisco · São Paulo · Sydney · Tokyo · Toronto

ACADEMIC PRESS INC. (LONDON) LTD.
24–28 Oval Road,
London NW1

United States Edition published by
ACADEMIC PRESS INC.
111 Fifth Avenue
New York, New York 10003

Blackett, D. W.
Elementary topology
1. Algebraic topology
I. Title
514'.2   QA612

ISBN 0-12-103060-1
LCCN 82-45023

PRINTED IN GREAT BRITAIN

PREFACE TO THE REVISED EDITION

Since the first edition went out of print, there have been many requests that this book be reissued. The immediate impetus for this revised edition was the interest expressed at The Open University.

This book, with minimal formal prerequisites, has been used in many ways. In addition to being a text for a course in the regular mathematics curriculum, it has been the basis of independent study by beginning students eager to supplement the mathematics in the prescribed courses and has been the text in extension courses for older students. For the student continuing to advanced mathematics, the book has been an appetizer for algebraic topology; for others, a dessert which complements prior study of geometry.

PREFACE

This book has been written as a text for a one-semester course in topology for sophomores, juniors, and seniors who have completed a year of calculus. For the prospective graduate student, the course is to stimulate interest in topology; for the prospective secondary school teacher, it is to broaden his mathematical perspective.

For all its importance in the mathematical world, topology has been slighted in the undergraduate curriculum. Many colleges and universities do not offer a topology course designed primarily for undergraduates. Those that do usually treat topology as a tool for analysis. While point-set methods are essential for modern analysis, a course limited to point-set topology may refine a student's analytic technique but kill his interest in topology as an independent discipline. To emphasize topological concepts and theorems which will be meaningful to the student, this book treats selected topics from one, two, and three dimensions by primarily combinatorial and algebraic methods. It is hoped that these topics will be exciting to both student and teacher.

Chapter 1 introduces the student to the sphere, the torus, the cylinder, the Möbius band, and the projective plane. In Chapter 2 combinatorial techniques are used to classify surfaces. Starting from the student's knowledge of conic sections in plane analytic geometry, Chapter 3 uses the methods of the preceding chapter to describe topologically the loci of quadratic equations in two complex variables. These special two-sheeted coverings of the Riemann sphere lead to a discussion of covering surfaces. The concept of winding number is employed in Chapter 4 to study mappings into the sphere. The results include the two-dimensional cases of the Brouwer fixed point theorem and the Borsuk-Ulam theorem. A further application is a proof of the fundamental theorem of algebra coupled with a method for isolating roots of complex polynomial equations. Chapter 5 studies vector fields on the plane and sphere. The discussion includes applications to cartography, hydrodynamics, and differential equations. In Chapter 6 the homology of networks is developed and applied to the Kirchhoff-Maxwell laws and to a transportation problem. The final chapter is a brief introduction to three-dimensional manifolds.

The exercises at the end of the chapters include drill problems, illustrations of the text, and extensions of the text. The text has many worked examples. In some sections the concepts are developed primarily through examples.

A teacher may elect to emphasize certain chapters and omit others. Chapters 3, 4, and 7 depend only on Chapters 1 and 2; Chapter 5 depends on Chapter 4; and Chapter 6 is independent.

A semester course based on this book might be preceded or followed by a semester course in point-set topology, projective geometry, or differential geometry to give a full year of geometry. The text might be used for a portion of an integrated year course in modern geometry. It might also be used in a summer institute for secondary school teachers.

Several years ago I started teaching a one-semester course patterned after the undergraduate topology course developed by Professor Albert W. Tucker at Princeton University. Professor Tucker generously allowed me to start from his lecture materials and exercises in writing this textbook to fit the course. The lecture notes, recorded by Mr. R. C. James when Professor Tucker presented his course as Phillips lecturer at Haverford College, have been especially helpful. Mr. James rewrote and published a portion of these notes as " Combinatorial Topology of Surfaces," *Mathematics Magazine*, Vol. 29 (1955) 1–39. A list of exercises compiled by Professor E. F. Whittlesey from Professor Tucker's course records has been a source of many exercises. I am grateful to Professor Herman R. Gluck for his many suggestions for improving the manuscript.

Boston, Massachusetts   D. W. B.

CONTENTS

3.   Complex Conics and Covering Surfaces

4.   Mappings into the Sphere

5.   Vector Fields

6.   Network Topology

7. Some Three-Dimensional Topology

Bibliography

Subject Index

ELEMENTARY TOPOLOGY
A Combinatorial and Algebraic Approach

1 SOME EXAMPLES OF SURFACES

1.1 Coordinates on a Sphere and Torus

Two of the simplest examples of surfaces in three-dimensional space are the *sphere*, which is the surface of a solid ball, and the *torus*, which is the surface of a doughnut. If x, y, z are rectangular coordinates in three-dimensional Euclidean space, the loci of the equations

$$x^2 + y^2 + z^2 = 1 \quad \text{and} \quad z^2 = (3 - \sqrt{x^2 + y^2})(\sqrt{x^2 + y^2} - 1)$$

are examples of a sphere and torus, respectively.†

To see that the second locus is a torus, consider the locus as the union of the graphs of the two functions of x and y defined by the equations

$$z = \sqrt{(3 - \sqrt{x^2 + y^2})(\sqrt{x^2 + y^2} - 1)},$$

$$z = -\sqrt{(3 - \sqrt{x^2 + y^2})(\sqrt{x^2 + y^2} - 1)}.$$

† In both verbal and graphic descriptions of geometric figures in space we adopt the following conventions: the positive direction of the x-axis is back to front, the positive direction of the y-axis is left to right, and the positive direction of the z-axis is down to up or, equivalently, south to north.

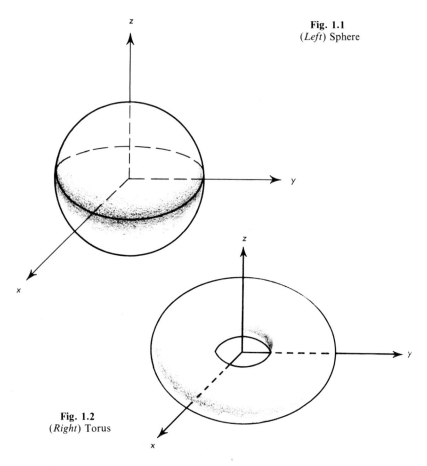

Fig. 1.1
(*Left*) Sphere

Fig. 1.2
(*Right*) Torus

If the point (x, y) is inside the circle $x^2 + y^2 = 1$, $\sqrt{x^2 + y^2} - 1$ is negative but $3 - \sqrt{x^2 + y^2}$ is positive. Hence neither formula gives a real number z, for no real number z could have a negative square. Similarly, if (x, y) is outside the circle $x^2 + y^2 = 9$, no values of z are defined, for $3 - \sqrt{x^2 + y^2}$ is negative, whereas $\sqrt{x^2 + y^2} - 1$ is positive. When (x, y) is between the circles so that $1 < x^2 + y^2 < 9$, the two formulas define two distinct values of z, one positive and one negative. If (x, y) is on the circle $x^2 + y^2 = 1$ or the circle $x^2 + y^2 = 9$, both formulas give the value $z = 0$. Thus the two functions defining z in terms of x and y have as domain the annulus determined by the inequalities

$$1 \le x^2 + y^2 \le 9,$$

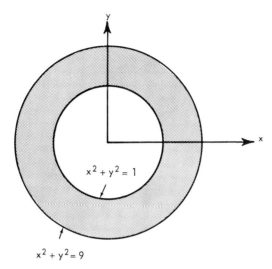

Fig. 1.3

and the two functions agree only on the two circles that bound this annulus.

The graph of the first function is the part of the locus of

$$z^2 = (3 - \sqrt{x^2 + y^2})(\sqrt{x^2 + y^2} - 1)$$

for which $z \geq 0$. The graph of the second function is the part of the locus for which $z \leq 0$. Thus the locus consists of two sheets, one above and one below the annulus $1 \leq x^2 + y^2 \leq 9$ in the plane $z = 0$. These sheets meet along the circles $x^2 + y^2 = 1$ and $x^2 + y^2 = 9$ in the plane $z = 0$ to form a ring-shaped surface. We shall see shortly that a vertical plane passing through the origin intersects this surface in a pair of circles. "Torus" is the mathematical name for a ring-shaped surface.

A coordinate system assigns names to points. To understand and solve different problems it is convenient to define various types of coordinate. Whenever two or more coordinate systems are used, we should know how to translate from one coordinate language to another. For example the *polar coordinates* (r, θ) in a plane with *rectangular coordinates* (x, y) are related to the rectangular coordinates by the equations

$$x = r \cos \theta, \qquad y = r \sin \theta.$$

If r and θ replace x and y in the coordinate system (x, y, z), the result is the *cylindrical coordinate system* (r, θ, z) for three-dimensional Euclidean space. The name is appropriate because the surface $r = c$ for c, a constant, is the infinite cylinder of points at a distance c from the z-axis.

Returning to our examples of a sphere and a torus, we find that their equations in cylindrical coordinates are

$$r^2 + z^2 = 1 \quad \text{and} \quad z^2 = (r - 1)(3 - r).$$

Because these equations impose no condition on θ, θ can assume any value. The angular coordinate θ of a point on the sphere or torus is the *longitude* of the point. Because a change in the angular cylindrical coordinate by an integral multiple of 2π radians gives a new set of coordinates for the same point, the longitude of a point has many values. We remove the ambiguity by requiring that $-\pi < \theta \leq \pi$.

When θ is the angular polar coordinate in the plane, the equation $\theta = c$, in which c is a constant, determines a halfline or ray emanating from the origin. If θ is the angular cylindrical coordinate, $\theta = c$ is the equation of a vertical halfplane which starts at the z-axis and extends in one direction. In this halfplane r and z are rectangular coordinates. By rewriting the equation of the torus as

$$(r - 2)^2 + z^2 = 1$$

we see that the intersection of the torus with the halfplane $\theta = c$ is a circle of radius 1 and with center two units away from the origin in a horizontal direction. Thus the particular torus we have been discussing is the torus of revolution generated by revolving the circle

$$(x - 2)^2 + z^2 = 1$$

in the xz-plane about the z-axis.

For points on the sphere $r^2 + z^2 = 1$, we define

$$\phi = \text{arc tan}\left(\frac{z}{r}\right).$$

(By arc tan t is meant the angle between $-\pi/2$ and $\pi/2$ radians whose tangent is t.) If $r = 0$, ϕ is $\pi/2$ when $z = 1$ and $-\pi/2$ when $z = -1$. The angle ϕ, called the *latitude*, is the angle of elevation of the point on the sphere viewed by an observer stationed at the center of the sphere. The geographical distinction between north latitude and south latitude is preserved if we interpret positive latitude as north latitude and negative latitude as south latitude. Because a point on the sphere is uniquely determined by its latitude and longitude, latitude and longitude form a coordinate system on the sphere. The two anomalies of the coordinate system are that latitude $\pi/2$, with any longitude, always names the north pole and that latitude $-\pi/2$, with any longitude, always specifies the south pole.

We shall now define latitude on our example of a torus. In the plane $z = 0$ the circle $r = 2$, which we call the *axial circle*, is the path of the center of the circle revolving about the z-axis to generate the torus. We define the latitude of a point on the torus as the angle of elevation ϕ of the point viewed by an observer facing away from the origin and stationed at the point on the axial circle with the same longitude as the given point on the torus (Figure 1.4). Algebraically,

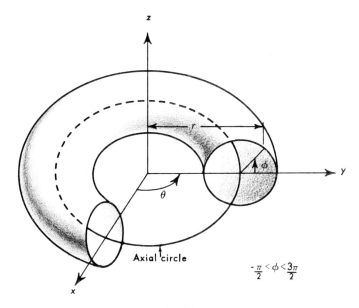

Fig. 1.4

$$\phi = \text{arc tan}\left(\frac{z}{r-2}\right) \quad \text{if } r > 2,$$

$$\phi = \frac{\pi}{2} \quad \text{if } r = 2 \text{ and } z = 1,$$

$$\phi = \pi + \text{arc tan}\left(\frac{z}{r-2}\right) \quad \text{if } r < 2,$$

$$\phi = \frac{3\pi}{2} \quad \text{if } r = 2 \text{ and } z = -1.$$

These formulas give ϕ as an angle in the first or second quadrant for all points on the "northern hemitorus" and as an angle in the third quadrant between π and $3\pi/2$ or in the fourth quadrant between 0 and $-\pi/2$ for all points on the "southern hemitorus." The latitude is greater than $\pi/2$ if the observer on the axial circle facing away from the origin must bend his head back an angle greater than $\pi/2$ to see a point behind him. As on the sphere, latitude and longitude form a coordinate system on the torus. The anomalies of the latitude-longitude system that occur at the north and south poles of the sphere do not appear on the torus.

Let us make a map of the torus by plotting latitude and longitude as rectangular coordinates in a plane (Figure 1.5). Because

$$-\pi < \theta \leq \pi \quad \text{and} \quad -\frac{\pi}{2} < \phi \leq \frac{3\pi}{2},$$

the map is a *rectangular area*† which includes the top and right edges but not the bottom or left edges. A heavy line in Figure 1.5 denotes an edge included in the map, whereas a light line indicates an edge not included. The solid dot represents a vertex included; a small circle stands for one not included. On this map each point corresponds to a unique point on the torus and each point on the torus has a unique image on the map.

If a point on the map approaches the missing left edge of the rectangle, the corresponding point on the torus approaches the meridian (curve of constant longitude) represented by the right edge of the map. Similarly, approaching the bottom edge of the map corresponds on the torus to approaching the parallel of latitude (curve of constant latitude) represented by the top edge of the map. These facts can be indicated on our map by adding the missing edges of the rectangle and explaining in the legend of the map that the $-\pi$- and π-meridians on the map represent the single π-meridian on the torus and that the $-\pi/2$ and $3\pi/2$ parallels on the map correspond to the single $3\pi/2$ parallel on the torus.

If we do not distinguish between the torus and the map of the torus, we can think of the torus as a rectangle with the opposite edges identified.

---

† In common English usage the words *rectangle*, *triangle*, and *polygon* can mean either an area or the curve that is the perimeter of the area. Because our primary concern is two-dimensional topology, we use these words to signify areas and use *rectangular curve*, *triangular curve*, and *polygonal curve* when the perimeter is meant. Our two-dimensional interest also dictates that *tetrahedron*, *cube*, and *polyhedron* mean surfaces, whereas *tetrahedral solid*, *cubical solid*, and *polyhedral solid* describe solids. The word *circle* means a curve bounding an area called a *disk;* the word *sphere* denotes the surface of a solid called a *ball*.

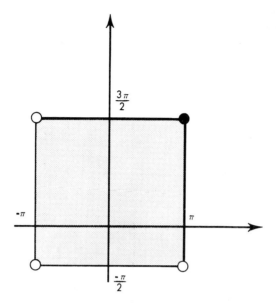

Fig. 1.5

Let a be the right edge directed upward and b the top edge directed from right to left. Figure 1.6 depicts a torus. The torus can be specified by the following listing of the edges in counterclockwise sequence around the perimeter of the rectangle (starting at the lower right corner):

$$aba^{-1}b^{-1} = 1.$$

The exponent -1 indicates that the direction of the edge is opposite to the direction of the sequence. The sequence is set equal to 1 as a notational device to show that the complete perimeter of the polygon has been given.

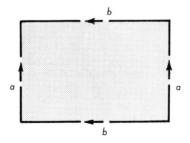

Fig. 1.6

If there were a geography on the torus, we could represent it on a rectangular map just as the geography of the (almost) spherical surface of the earth is depicted on rectangular maps. Figure 1.7 shows the geography of

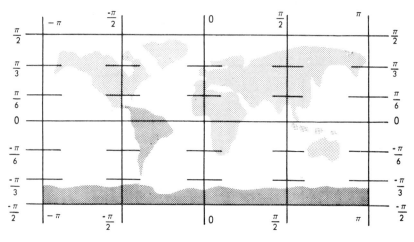

Fig. 1.7 Rectangular map of the earth's sphere

the earth on a map with latitude and longitude as rectangular coordinates. The $-\pi$- and π-meridians on the map correspond to a single meridian. The entire top edge of the map corresponds to the single point (the north pole) with latitude $\pi/2$. Similarly, the bottom edge represents the south pole.

To avoid having a whole line segment correspond to a single point, we make a new map of the sphere, using a pseudolongitude in place of the longitude. We define pseudolongitude ψ by

$$\psi = \left(1 - \frac{2|\phi|}{\pi}\right)\theta.$$

At latitude ϕ the pseudolongitude is restricted to the interval

$$-(\pi - 2|\phi|) \le \psi \le \pi - 2|\phi|.$$

Thus the length of the interval of pseudolongitude shrinks to zero as $|\phi|$ approaches $\pi/2$. In particular, the pseudolongitude of the north and south poles is zero. If pseudolongitude and latitude are plotted as rectangular coordinates, the resulting map of the sphere is a rhombus with vertical and horizontal diagonals (Figure 1.8). On this map the north and south poles are represented by unique points. Thus a sphere may be represented

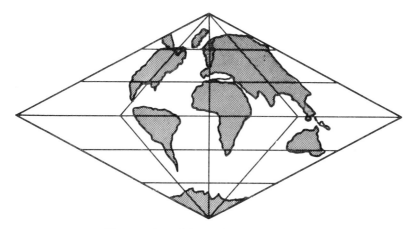

Fig. 1.8 Map of the earth as a rhombus

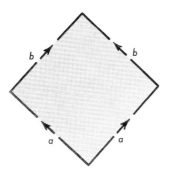

Fig. 1.9

as a quadrilateral with two pairs of adjacent edges identified (Figure 1.9). An edge sequence equation for the sphere is $abb^{-1}a^{-1} = 1$.

Stereographic projection of the sphere defines another planar map of the sphere. Through a point P on the sphere $r^2 + z^2 = 1$ draw the line determined by P and the north pole N. The point P' at which this line intersects the equatorial plane ($z = 0$) is the stereographic image of P. The north pole is the only point on the sphere with no image in the plane. Let ϕ and θ be the latitude and longitude of P. The angular polar coordinate of P' in the equatorial plane is also θ. We now calculate the polar coordinate r of P' in terms of the latitude of P (Figure 1.10).

The circle in Figure 1.10 is the cross section of the sphere in the plane determined by P, N, and the origin O. This plane is the set of points for

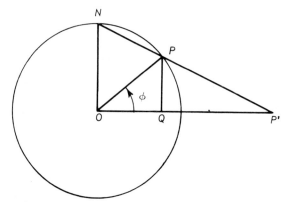

Fig. 1.10

which the angular cylindrical coordinate either equals the longitude of P or differs by π from this longitude. Because the line through P and N is in this plane, P' is also in the plane. The points P and P' are both below N, hence are both on the same side of N. Because they are on the same side of the z-axis, we have confirmed the earlier statement that P and P' have the same angular cylindrical coordinate. The point Q is the foot of the perpendicular from P to OP'. Because the radius of the sphere is 1, the length of OQ is $\cos\phi$ and of PQ, $\sin\phi$. Now the triangles NOP' and PQP' are similar. Hence

$$\frac{\sin\phi}{1} = \frac{r - \cos\phi}{r}$$

or

$$r = \frac{\cos\phi}{1 - \sin\phi}. \tag{1}$$

The figure and proof are appropriate for P in the northern hemisphere. If P is in the southern hemisphere, the result is the same and the proof similar. (See Exercise 1.7.) The rectangular coordinates of P' are

$$x = \frac{\cos\phi\cos\theta}{1 - \sin\phi}, \qquad y = \frac{\cos\phi\sin\theta}{1 - \sin\phi}.$$

For the points in a plane the rectangular coordinate pair (x, y) of real numbers may be replaced by the single complex number $w = x + iy$. (The letter z is more commonly used to denote $x + iy$, but we are currently using z as the third rectangular coordinate in space.) If x and y are expressed in terms of polar coordinates, we have the polar form of the complex number

$$w = r(\cos \theta + i \sin \theta).$$

The number r is the absolute value $|w|$, and θ is the argument of w, written arg w. If the equatorial plane is considered as the plane of complex numbers, stereographic projection maps the point P on the sphere with latitude ϕ and longitude θ onto the complex number

$$\frac{\cos \phi}{1 - \sin \phi} (\cos \theta + i \sin \theta).$$

In an alternative interpretation this complex number is the complex coordinate of the point P on the sphere. Every point except N has a complex coordinate. As P approaches N, the latitude approaches $\pi/2$. Now

$$r = \frac{\cos \phi}{1 - \sin \phi} = \frac{\cos \phi (1 + \sin \phi)}{1 - \sin^2 \phi} = \frac{1 + \sin \phi}{\cos \phi}.$$

As ϕ approaches $\pi/2$, r becomes arbitrarily large and P' moves arbitrarily far away from the origin. The missing point N' corresponding to N is supplied by adding to the plane of complex numbers a single point at infinity, written ∞. This new complex number is the complex coordinate of the north pole N. The complex plane with ∞ adjoined is the extended complex plane. When complex coordinates are used on the sphere, it is called the Riemann sphere in honor of G. F. Bernhard Riemann (1826–1866), who presented a geometric development of the theory of functions whose domain and range are sets of complex numbers.

Figure 1.11 is a stereographic map of the portion of the earth south of latitude $\pi/4$. It resembles the view of the southern hemisphere " seen " by an observer looking through the earth from the north pole. Stereographic projection has the useful property that two curves on the sphere intersecting at P with angle β are mapped into two curves in the plane intersecting at P' with angle β. A map that preserved angles is called *conformal*. As a consequence of conformality, the shape of a small area on a stereographic map is an approximation of the shape of the corresponding area on the sphere. The distorted shape of Africa in Figure 1.11 shows graphically that the local accuracy of a stereographic map does not imply global accuracy. Because of its precision throughout the south polar region, the stereographic projection we have described is frequently used to make maps of Antarctica.

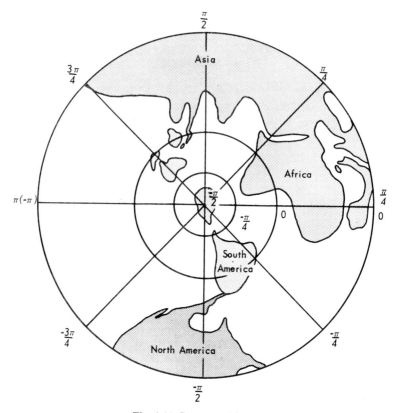

Fig. 1.11 Stereographic map

1.2 The Topological Sphere and Torus

In Section 1.1 we discussed a particular sphere and a particular torus in three-dimensional Euclidean space. Now we define the *topological sphere* and *topological torus*.

In order to do so we need the notion of *homeomorphism*. Suppose that A and B are two point sets in Euclidean spaces† and that a function

$$f : A \to B$$

---

† By a *Euclidean space* (finite-dimensional) we mean a set of points, $P, Q \dots$, with a distance function $d(P, Q)$ such that for some positive integer n the points may be uniquely represented by the n-tuples $(x_1, \dots, x_n)$ of real numbers and the distance between $P = (x_1, \dots, x_n)$ and $Q = (y_1, \dots, y_n)$ is

$$d(P, Q) = \sqrt{(x_1 - y_1)^2 + \cdots + (x_n - y_n)^2}.$$

with domain A and range B is

(a) *one-to-one*, that is, if a and a' are points of A such that $f(a) = f(a')$, then $a = a'$;

(b) *onto*, that is, for each point b in B there is some a in A such that $f(a) = b$.

Thus the inverse function

$$f^{-1}: B \to A$$

is also well-defined. Suppose, furthermore, that both f and f^{-1} are continuous. We then say that the function f (and, equally well, f^{-1}) is a *homeomorphism* and that the point sets A and B are *homeomorphic* or *topologically equivalent*. The adjective "topological" is used to describe properties which, if possessed by a given point set, must also be possessed by all point sets homeomorphic to the given one; for example, the intuitive notion of "connectedness" can be given a precise definition, and when this is done, connectedness turns out to be a topological property. On the other hand, the metric notion of "size" or "diameter" of a point set is not a topological property. To illustrate, let C_1 denote a circle of radius 1 and center at the origin in the plane; that is, C_1 is the locus of all points (x, y) in the plane which satisfy the equation

$$x^2 + y^2 = 1.$$

Let C_2 be the circle of radius 2 and center at the origin—that is, the locus of the equation

$$x^2 + y^2 = 4.$$

The function $f: C_1 \to C_2$ defined by the equation

$$f(x, y) = (2x, 2y)$$

satisfies all the properties of a homeomorphism (the reader should check this statement carefully). Thus C_1 and C_2 are homeomorphic, even though they have different diameters. With a little care the reader can show that any two circles are homeomorphic and also that any circle is homeomorphic to any square. On the other hand, a circle and a figure eight are not homeomorphic. It is a little harder to show that two point sets are not homeomorphic because it is, of course, not sufficient to show that a particular map between them is not a homeomorphism—in effect we must show that every possible map between them is not a homeomorphism.

The real numbers $x_1, \ldots, x_n$ are called *rectangular* or *Cartesian coordinates* of P. Since Euclidean spaces are adequate to illustrate the concepts and theorems of this book, the temptation to consider more general types of space has been resisted.

Intuitively, the circle is not homeomorphic to the figure eight because the figure eight can be disconnected by the removal of one point (the crossing point), but the circle cannot be disconnected by the removal of any single point (two points would have to be removed from a circle to disconnect it). A precise argument can be built up around this intuitive one, once the notion of connectedness has been defined. Similarly, a circle and a line segment are not homeomorphic.

If we examine the definition of homeomorphism carefully, it can be seen that a half-open line segment (a segment with one end point and missing the other) comes very close to being homeomorphic to a circle. Consider the line segment $[0, 1)$ consisting of all numbers t such that $0 \le t < 1$. Consider the map

$$f:[0, 1) \to C_1,$$

given by the formula

$$f(t) = (\cos 2\pi t, \sin 2\pi t).$$

Now f is one-to-one, onto, and continuous, but the map

$$f^{-1}:C_1 \to [0, 1)$$

fails to be continuous at the point $(1, 0)$. This shows how important it is in the definition of homeomorphism to require that both f and f^{-1} be continuous, for it is clear to the eye that a circle and a line segment are grossly different.

Most of the sets we shall study are closed and bounded. It is a simple theorem of point set topology that if f is a one-to-one continuous mapping of a closed bounded set A onto a set B, then B is closed and bounded and f^{-1} is continuous. Thus in this special case the continuity of f^{-1} need not be explicitly verified.

Now we are ready to define the topological torus.

Definition. A point set T in some Euclidean space is called a *topological torus* if T is homeomorphic to the particular torus of Section 1.1.

The adjective " topological " is dropped when the meaning is clear from context.

Consider again the particular torus of Section 1.1. The correspondence between points (θ, ϕ) in the $\theta\phi$-plane and points on this torus defines a continuous function on a quadrilateral in this plane with values on the torus. Every point on the torus appears once and only once as the image of a point in the quadrilateral, except that the π-meridian and the $-\pi$-meridian on the quadrilateral map into a single meridian on the torus and the $3\pi/2$ and $-\pi/2$ parallels on the quadrilateral have a single parallel on the torus as their image.

Similarly, if a point set T in a Euclidean space is a topological torus, it is the image of a plane quadrilateral Σ under a continuous function f with the following properties:

1. If P is not on the perimeter of Σ, then $f(P) \neq f(Q)$ for any $Q \neq P$ in Σ.

2. If P is on the perimeter of Σ but is not a vertex, there is exactly one point P' different from P such that $f(P) = f(P')$. P and P' are on opposite edges of Σ. As P moves along an edge in the counterclockwise direction around Σ, P' moves in a clockwise direction.

3. If P_1, P_2, P_3, P_4 are the four vertices of Σ,

$$f(P_1) = f(P_2) = f(P_3) = f(P_4).$$

The quadrilateral with opposite edges identified as indicated by the equation $aba^{-1}b^{-1} = 1$ is considered a generic torus. The function f specifies a particular way of embedding the generic torus in a Euclidean space to define a particular torus.

We shall now give an example of a torus in a four-dimensional Euclidean space with rectangular coordinates x_1, x_2, x_3, x_4. The torus is the locus of the parametric equations

$$x_1 = \cos u, \qquad x_2 = \cos v, \qquad x_3 = \sin u, \qquad x_4 = \sin v,$$

with $0 \leq u \leq 2\pi$ and $0 \leq v \leq 2\pi$. The restrictions on the parameters u and v determine a square in the uv-plane. The embedding function f is defined by

$$f(u, v) = (\cos u, \cos v, \sin u, \sin v).$$

Now $f(0, v) = f(2\pi, v)$ and $f(u, 0) = f(u, 2\pi)$. Because there are no other circumstances under which $f(u_1, v_1) = f(u_2, v_2)$, unless $u_1 = u_2$ and $v_1 = v_2$, this locus in four-dimensional Euclidean space is a topological torus. Naturally it is difficult to visualize this particular torus.

Now that we have a topological definition of the torus we shall consider the sphere.

Definition. A point set S in some Euclidean space is called a *topological sphere* if it is homeomorphic to the particular sphere of Section 1.1.

Equivalently, a topological sphere is a point set on which latitude and longitude, latitude and pseudolongitude, or a complex coordinate can be defined to satisfy the same identification properties these coordinates satisfy on the particular sphere $x^2 + y^2 + z^2 = 1$. As an example, we show that a cube is a topological sphere. First inscribe the cube in a Euclidean sphere. Through any point P on the cube draw a radius of the sphere. This radius intersects the sphere in a point P' (Figure 1.12). The map from the cube to the sphere which sends P onto P' is a homeomorphism. Thus the cube is homeomorphic to some Euclidean (round) sphere. As in the case of the circles, all spheres are homeomorphic to one another; therefore the cube is indeed a topological sphere. Alternately, if the latitude and longitude of P are defined to be the latitude and longitude of P', the latitude and longitude system on the cube has the same identification properties as that on the sphere, and again the cube is seen to be a topological sphere.

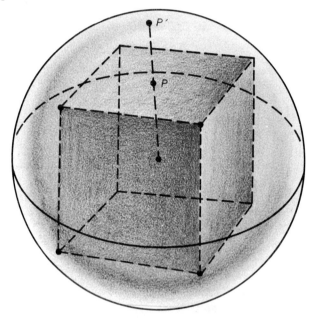

Fig. 1.12

The cube is an example of an extensive class of surfaces that are topologically spheres. Stretch, bend, compress, but do not tear a Euclidean sphere. If the latitude and longitude on the deformed surface are defined as the latitude and longitude before the deformation, the continuity of the permitted operations ensures that this definition of coordinates on the deformed surfaces establishes that the deformed surface is topologically a sphere. The cube may be formed by compressing the circumscribing Euclidean sphere.

As a different type of example of a sphere, consider the locus in four-dimensional Euclidean space defined by the parametric equations

$$\left.\begin{array}{l} x_1 = u(1 - |u|)(1 - |v|) \\ x_2 = v(1 - |u|)(1 - |v|) \\ x_3 = (1 - |u|)(1 - |v|) \\ x_4 = u + v \end{array}\right\} \begin{array}{l} -1 \le u \le 1 \\ -1 \le v \le 1 \end{array}.$$

The function

$$f(u, v) = (u(1 - |u|)(1 - |v|), v(1 - |u|)(1 - |v|), (1 - |u|)(1 - |v|), u + v)$$

defined over the square $\max(|u|, |v|) \le 1$ has each point on the locus as the image of a point in the square. Now, the third coordinate x_3 of $f(u, v)$ is not zero unless (u, v) is on the perimeter of the square. Because

$$u = \frac{x_1}{x_3} \quad \text{and} \quad v = \frac{x_2}{x_3} \quad \text{when } x_3 \ne 0,$$

$f(u_1, v_1) \ne f(u_2, v_2)$ unless $u_1 = u_2$ and $v_1 = v_2$ if (u_1, v_1) is not on the perimeter of the square. For points on the perimeter the identities

$$f(1, t) = f(t, 1) \qquad f(-1, t) = f(t, -1)$$

exhibit the only pairs of points on the perimeter with identical images on the locus. This is because the line $u + v = x_4$ (considering x_4 as a constant) intersects the perimeter of the square either in points of the form $(1, t)$ and $(t, 1)$ or in points of the form $(-1, t)$, $(t, -1)$. Thus the four-dimensional locus is a realization of the surface formed from a quadrilateral by the identification of adjacent edges as specified by the edge equation $abb^{-1}a^{-1} = 1$. This is the identification pattern of the latitude-pseudo-longitude coordinate system on the sphere. We now recognize the four-dimensional locus as a topological sphere.

We have discussed the sphere rather intuitively. The sphere could have been defined in terms of properties of an embedding function.

Another example of a sphere is the locus of the equation $w^2 = z$ in the two-dimensional complex space of pairs (z, w). The coordinates $z = x + iy$ and $w = u + iv$ range over the extended complex number system so that the point (∞, ∞) may be included in the locus. The function that assigns to each complex number w (including ∞) the point (w^2, w) on the locus is a homeomorphism. Equivalently w may be used as a single complex coordinate for the point (w^2, w). Thus the locus is topologically a sphere. In contrast we shall discover in Chapter 3 that the locus of

$$w^2 = z(z + 1)(z - 1)$$

is a torus. Because the complex pair (z, w) is equivalent to the real quadruple (x, y, u, v), four-dimensional perception is necessary to see these loci in their natural habitat. The algebraic methods of Chapter 3 will enable us to classify such loci topologically without the eyestrain of four-dimensional vision.

1.3   Properties of the Sphere and Torus

Now that we have given topological definitions of the sphere and torus, we wish to study the properties common to all particular spheres or tori that satisfy the topological definitions. Among these topological properties are many that distinguish between the sphere and the torus.

In considering convex polyhedra† Leonhard Euler (1707–1783) discovered a remarkable fact. If the number N_1 of edges of the polyhedron is subtracted from the sum of the number N_0 of vertices and the number N_2 of faces, the answer is always two. This is proved in Chapter 2. Examples of convex polyhedra are the five regular polyhedra: the tetrahedron, the cube, the octahedron, the dodecahedron, and the icosahedron (Figure 1.13).

	N_0	N_1	N_2	$N_0 - N_1 + N_2$
Tetrahedron	4	6	4	2
Cube	8	12	6	2
Octahedron	6	12	8	2
Dodecahedron	20	30	12	2
Icosahedron	12	30	20	2

A polyhedron is *convex* under the following conditions:

1. It divides space into two nonempty parts, an unbounded one called the *outside* and a bounded one called the *inside*.

2. Any line L passing through the inside cuts the polyhedron at exactly two points P_1 and P_2 so that all points on L between P_1 and P_2 are inside, whereas all other points on L except P_1 and P_2 are outside.

Let O be a particular point inside a convex polyhedron. Any line through O intersects the polyhedron in two points P_1 and P_2 separated by O. Leaving O fixed, shrink or stretch the line segments OP_1 and OP_2 until P_1P_2 is a line segment of length 2 with O as midpoint. This defines a continuous mapping of the polyhedron onto a sphere. By assigning to each point on the polyhedron the latitude and longitude of its image on the sphere we establish that the convex polyhedron is topologically a sphere. A topological statement of Euler's discovery is that if a sphere is divided

---

† A precise definition of "polyhedron" is given in Chapter 2.

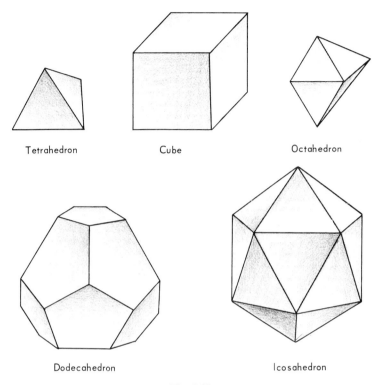

Fig. 1.13
Convex polyhedra

by curves into a finite number of faces (each bounded by a single curve), the number N_1 of edges (the arcs between intersection points of the curves) subtracted from the sum of the number N_0 of vertices (the intersection points of the curves) and the number of N_2 of faces is always two.

The surface of a picture frame (Figure 1.15) is a nonconvex polyhedron which is topologically a torus. This polyhedron has 16 vertices, 32 edges, and 16 faces. For this or any other subdivision of a torus into a finite number of faces

$$N_0 - N_1 + N_2 = 0.$$

More generally, for any subdivision of a surface into N_0 vertices, N_1 edges, and N_2 faces, the number $N_0 - N_1 + N_2$ depends only on the surface and not on the subdivision. This number is the *Euler characteristic* of the surface. Because the Euler characteristic of the sphere is two and that of the

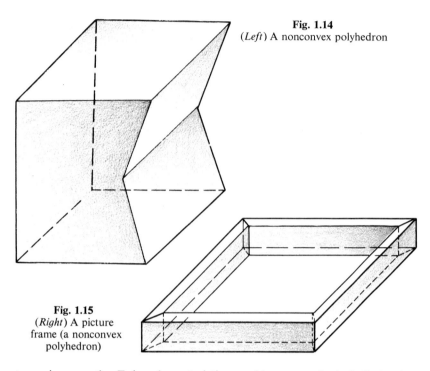

Fig. 1.14
(*Left*) A nonconvex polyhedron

Fig. 1.15
(*Right*) A picture
frame (a nonconvex
polyhedron)

torus is zero, the Euler characteristic provides a topological distinction between the sphere and torus.

The Euler characteristic can be used to derive other properties of a sphere or torus. Let a sphere or torus be divided by curves into a finite number of faces which we call *countries*. An edge common to two countries is a *frontier* between them. The subdivision of the sphere or torus is a *map*. A coloring of the map assigns a color to each country so that two countries sharing a common frontier are assigned different colors. Note that two countries that meet only at a vertex, as do the states of Colorado and Arizona, may be assigned the same color. We now use the Euler characteristic to show that every map on the sphere can be colored with six or fewer colors but that some maps on the torus require seven colors.

Suppose there are maps on a surface that require seven or more colors. Among all such maps select one with the smallest number of countries. We now establish that every country on this minimal map must have at least six frontiers. If there is a country with fewer than six frontiers, annex it to one of its neighbors. Because the resulting map has one less country, the minimal map after annexation can be colored with six colors. Select

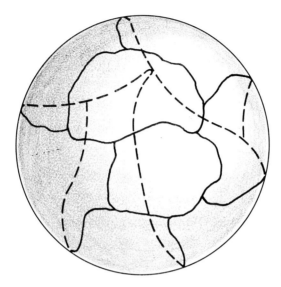

Fig. 1.16
A map on a sphere with $N_0 = 11$, $N_1 = 17$, $N_2 = 8$

such a coloring and let the annexed country regain its independence. Since the liberated country has common frontiers with no more than five other countries, its color as a captive can be replaced by one of the six permissible colors which has not been assigned to any of the neighboring countries. This contradiction of the assumption that the minimal map cannot be colored with six colors was established by finding a country with fewer than six frontiers. Hence every country on the minimal map has six or more frontiers. If each country assigns one guard to each of its frontiers, the number of guards is at least $6N_2$. Because every frontier is common to exactly two countries, the number of guards is $2N_1$. Hence

$$2N_1 \geq 6N_2 .$$

Each frontier has one vertex at each end and each vertex is at the end of three or more frontiers. If, at every vertex, there is a signpost pointing along each frontier, the number of signposts equals $2N_1$ and is greater than or equal to $3N_0$. Hence

$$2N_1 \geq 3N_0 .$$

Therefore

$$N_0 - N_1 + N_2 \leq \frac{2}{3} N_1 - N_1 + \frac{1}{3} N_1 = 0.$$

This shows that no map requires more than six colors on a surface, such as a sphere, with positive Euler characteristic.

Figure 1.17 depicts a map on a torus that requires seven colors. The map has seven hexagonal countries, each pair of which shares a common frontier. Similar reasoning establishes that every map on the torus can be colored with seven or fewer colors. A refinement of the proof shows that every map on the sphere can be colored with five or fewer colors. (See pp. 246–248, *What is Mathematics*, Courant and Robbins, Oxford University press, 1941.) In 1976 Appel and Haken announced a computer-assisted proof that every map can be colored with four colors. (See K. Appel and W. Haken, "Every Planar Map is Four Colourable", *Bull. A. M. S.* **82** (1976) pp. 711–712.)

Another application of the Euler characteristic relates to the problems of finding regular subdivisions. A *subdivision* of a surface into a polyhedron is *regular* if each face has the same number of edges and each vertex has

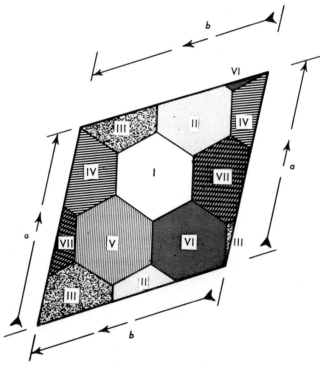

Fig. 1.17 Seven-color map on a torus

the same *order*. The order of a vertex is the number of times the vertex appears as the end of an edge. If the faces of a regular subdivision have k edges and the vertices are each of order j,

$$2N_1 = kN_2 \quad \text{and} \quad 2N_1 = jN_0.$$

These equations are derived by counting the face-edge and vertex-edge combinations represented by guards and signposts in the map-coloring problem. We find

$$N_0 - N_1 + N_2 = \left(\frac{2}{j} - 1 + \frac{2}{k}\right)N_1.$$

If the surface is a sphere,

$$\left(\frac{2}{j} - 1 + \frac{2}{k}\right)N_1 = 2.$$

If the surface is a torus,

$$\frac{2}{j} - 1 + \frac{2}{k} = 0.$$

To look for regular subdivisions of the sphere we rewrite the equation as

$$\frac{1}{j} + \frac{1}{k} - \frac{1}{N_1} = \frac{1}{2}.$$

Once j, k, and N_1 are found, N_0 and N_2 are also determined. Because the sum of the two positive terms on the left must be greater than $\frac{1}{2}$, at least one of the terms $1/j$ and $1/k$ must exceed $\frac{1}{4}$. Hence either j or k is less than 4. If $j = 3$ and $k \geq 6$ or $j \geq 6$ and $k = 3$,

$$\left(\frac{1}{j} + \frac{1}{k} - \frac{1}{N_1}\right) \leq \left(\frac{1}{3} + \frac{1}{6} - \frac{1}{N_1}\right) < \frac{1}{2}.$$

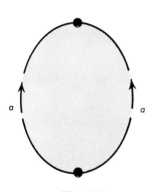

Fig. 1.18

Case 1. $j = 1$. From $1/N_1 - 1/k = 1/2$ we find

$$N_1 = 1, \quad k = 2, \quad N_2 = 1, \quad N_0 = 2.$$

This solution is realized if an arc is drawn on the sphere. The two ends of the arc are vertices of order 1 and the single edge appears twice as an edge of the rest of the sphere considered as a two-sided polygon. The sphere may be represented as a two-sided polygon with the edge identification equation $aa^{-1} = 1$ (Figure 1.18).

Case 2. $j = 2$. From $1/k - 1/N_1 = 0$ it follows that $k = N_1$, $N_2 = 2$, $N_0 = N_1$. To see that this solution actually occurs, draw the equator on a Euclidean sphere and divide the equator into N_1 arcs. The northern and southern hemispheres are two faces, with N_1 edges and N_1 vertices, each of which has order 2 (Figure 1.19).

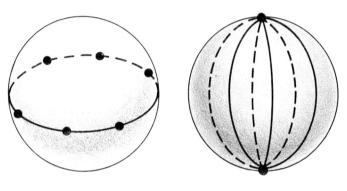

Fig. 1.19   **Fig. 1.20**

Case 3. $k = 1$. By interchanging j and k in Case 1 we derive $N_1 = 1$, $j = 2$, $N_2 = 2$, and $N_0 = 1$. This solution is included in Case 2.

Case 4. $k = 2$. By symmetry with Case 2, $j = N_1$, $N_2 = N_1$, and $N_0 = 2$. This solution is illustrated by drawing N_1 meridians on a Euclidean sphere. The north and south poles are two vertices of order N_1 and the spherical sectors between the meridians are N_1 two-sided polygons. Case 1 is included in Case 4 (Figure 1.20).

Case 5. $j = 3, k = 3$. From $1/3 + 1/3 - 1/N_1 = 1/2$ we find that $N_1 = 6$, $N_2 = 4$, and $N_0 = 4$. The tetrahedron is a regular polyhedron with six edges, four triangles, and four vertices, each of order 3.

Case 6. $j = 3, k = 4$. From $1/3 + 1/4 - 1/N_1 = 1/2$ it follows that $N_1 = 12$, $N_2 = 6$, and $N_0 = 8$. The cube meets this prescription.

Case 7. $j = 4, k = 3$. By symmetry with Case 6, $N_1 = 12$, $N_2 = 8$, and $N_0 = 6$. The regular octahedron satisfies these specifications.

Case 8. $j = 3, k = 5$. From $1/3 + 1/5 - 1/N_1 = 1/2$ we derive $N_1 = 30$, $N_2 = 12$, and $N_0 = 20$. These values fit the regular dodecahedron.

Case 9. $j = 5, k = 3$. The regular icosahedron with $N_0 = 12$, $N_1 = 30$, and $N_2 = 20$ is a realization of this case.

We have now considered all types of regular subdivision of the sphere.

For a regular subdivision of the torus the positive integers j and k satisfy the equation

$$\frac{1}{j} + \frac{1}{k} = \frac{1}{2}.$$

Either j or k must be less than or equal to 4 and both are greater than 2. One solution is $j = 3$, $k = 6$. For this solution $N_0 = 2N_2$ and $N_1 = 3N_2$. For every positive integer N_2 there is a subdivision of the torus into N_2 hexagons with $3N_2$ edges and $2N_2$ vertices, each of order 3. The map on the torus with seven hexagonal countries, shown in Figure 1.17, is an example of $N_2 = 7$. To give an example in which $N_2 = 1$, consider a single hexagon with opposite edges identified as in Figure 1.21 and in the equation

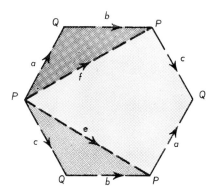

Fig. 1.21

$abca^{-1}b^{-1}c^{-1} = 1$. Cut this hexagon along the dotted lines e and f and reassemble the pieces to perform the actual identification of the two occurrences of a, the two occurrences of b, and the two occurrences of c. Because the surface is a quadrilateral (Figure 1.22) with the edge equation $efe^{-1}f^{-1} = 1$, we find that the original hexagon with opposite edges identified is a torus. The subdivision has one hexagon, three distinct edges, and two distinct vertices P and Q, each of order 3.

Figure 1.23 shows that the edges of seven hexagons in a row can be identified so that the surface formed is a torus. The fact that each pair of hexagons in Figure 1.17 shares a common edge but does not in Figure 1.23 demonstrates that a torus has essentially different regular hexagonal subdivisions with the same number of hexagons.

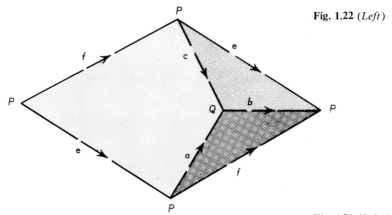

Fig. 1.22 (*Left*)

Fig. 1.23 (*Below*)

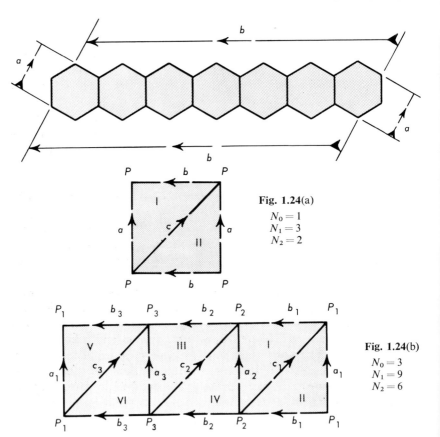

Fig. 1.24(a)
$N_0 = 1$
$N_1 = 3$
$N_2 = 2$

Fig. 1.24(b)
$N_0 = 3$
$N_1 = 9$
$N_2 = 6$

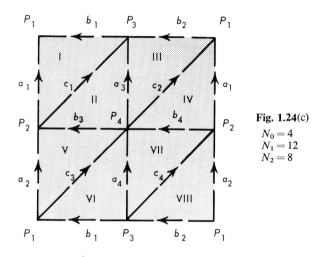

Fig. 1.24(c)
$N_0 = 4$
$N_1 = 12$
$N_2 = 8$

A second solution for j and k is $j = 6, k = 3$. For this solution $N_1 = 3N_0$ and $N_2 = 2N_0$. For every positive integer N_0 there is a subdivision of the torus into $2N_0$ triangles, $3N_0$ edges and N_0 vertices, each of order 6. Figure 1.24 illustrates subdivisions with $N_0 = 1, 3, 4$.

The third and last solution for j and k is $j = 4, k = 4$. For this solution $N_0 = N_2$, $N_1 = 2N_2$. For every positive integer N_2 there is a regular subdivision of the torus into quadrilaterals. Figures 1.25, 1.26, and 1.27 show subdivisions with $N_2 = 1, 3, 4$.

Using the Euler characteristic, we have found certain topological differences between the sphere and the torus. Every map on the sphere can be colored with five or fewer colors, whereas some maps on the torus require seven colors. A sphere can be regularly subdivided into pentagons; a torus cannot. A torus can be regularly subdivided into an arbitrarily large number of triangles, quadrilaterals, or hexagons, whereas a sphere has no such subdivisions with more than twenty faces.

We conclude this section with an example that illustrates the topological properties of the sphere and torus to be studied in Chapter 5. On the torus the meridians or the parallels form a family of closed curves such that each point of the torus is on one and only one curve of the family. In Chapter 5 we shall find that no such family of curves can be drawn on a sphere. In particular, the parallels on a sphere are not a suitable family because the north and south poles are on no parallel. The meridians are unsatisfactory because all meridians, hence more than one, pass through the north and south poles.

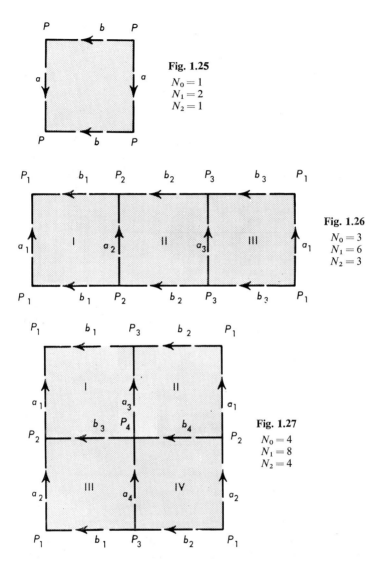

Fig. 1.25
$N_0 = 1$
$N_1 = 2$
$N_2 = 1$

Fig. 1.26
$N_0 = 3$
$N_1 = 6$
$N_2 = 3$

Fig. 1.27
$N_0 = 4$
$N_1 = 8$
$N_2 = 4$

1.4   The Cylinder and Möbius Band

An example of the *cylinder* is the set of points in three-dimensional Euclidean space whose cylindrical coordinates satisfy the conditions $r = 2$ and $|z| \leq 1$. The angular coordinate θ and the altitude z form a longitude and altitude system for the cylinder. If these coordinates are plotted as rectangular coordinates in a plane, the cylinder will have a rectangular map with $-\pi \leq \theta \leq \pi$ and $-1 \leq z \leq 1$, with each pair of points (z, π) and $(z, -\pi)$ considered as a single point. Thus the cylinder is represented as a quadrilateral with the edge equation $aba^{-1}c = 1$. Any point set in a Euclidean space with such a representation is topologically a cylinder. In a plane the annulus defined by the polar inequalities $1 \leq r \leq 2$ is topologic-

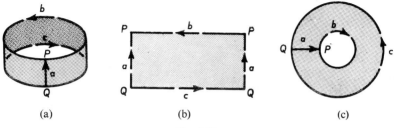

Fig. 1.28
(a) Cylinder, (b) $aba^{-1}c = 1$, (c) Annulus

ally a cylinder. If we use the polar coordinates (r, θ) with $-\pi \leq \theta \leq \pi$ for points in the annulus, (r, π) and $(r, -\pi)$ will represent the same points. This identification agrees with the equation $aba^{-1}c = 1$. When the cylinder is attached to another surface† along the edge c, the cylinder is called a *cuff*.

The edges b and c corresponding to the extreme values of the altitude can be approached on the surface from only one side. In contrast, the edge a appearing twice on the map and twice in the edge equation can be approached from both sides. The unmatched edges b and c are called *boundary edges*, whereas the matched edge a is an *interior edge*. The boundary edges are grouped together into simple closed curves‡ called *boundary curves*. Since b and c are both closed curves, each is a boundary curve.

Our particular cylinder $r = 2$, $|z| \leq 1$ is inscribed in the torus $(r - 2)^2 + z^2 = 1$ so that the boundary curves of the cylinder are on the torus. The toroidal solid defined by $(r - 2)^2 + z^2 \leq 1$ is divided into two parts by the cylinder (Figure 1.29). One part consists of points with $r < 2$ and the other of points with $r > 2$. If the two parts are interpreted as thick coats of paint, one red and one blue, the cylinder has been painted so that two distinguishable sides are different colors.

---

† The first section of Chapter 2 is a detailed explanation and definition of "surface."
‡ The points of a curve are the values of a continuous function f defined on the interval $0 \leq t \leq 1$. The initial and terminal vertices of the curve are the points $f(0)$ and $f(1)$, respectively. If $f(0)$ and $f(1)$ coincide, the curve is *closed*. If $f(t) \neq f(s)$, unless $t = s$ or $|t - s| = 1$, the curve does not intersect itself and is said to be *simple*.

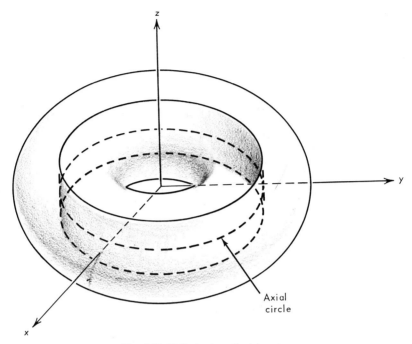

Fig. 1.29 Cylinder inscribed in a torus

We now describe another surface inscribed in the same torus. This surface is the locus specified by

$$(r-2)^2 + z^2 \leq 1 \quad \text{and} \quad z \sin\left(\frac{\theta}{2}\right) = (r-2)\cos\left(\frac{\theta}{2}\right).$$

The cross section of this surface at longitude θ is a diameter of the circle which is the cross section of the torus. This diameter rises in the direction away from the origin with slope $\cot(\theta/2)$. When $\theta = 0$, the diameter is vertical. As θ increases from 0 to π, the diameter rises less and less steeply until the diameter is in the horizontal plane $z = 0$ when $\theta = \pi$. If $-\pi < \theta < 0$, $\cot \theta/2$ is negative, and the diameter falls in the direction away from the origin. As θ increases from $-\pi$ to 0, the diameter at first is horizontal and then falls more and more steeply until the diameter is vertical when $\theta = 0$. This surface is an example of a *Möbius band*. A. F. Möbius (1790–1868), a German mathematician, was a pioneer topologist.

If P is a point on this Möbius band, let t be a number whose absolute value is the distance of P from the axial circle of the torus. Let t be positive

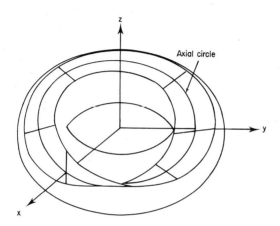

Fig. 1.30 Möbius band inscribed in a torus

when P is above the plane $z = 0$ and negative when P is below this plane. This definition is ambiguous at longitude π, for there the cross section is in the plane $z = 0$. The only other points on the surface in the plane $z = 0$ are those of the axial circle, where $t = 0$. If the point P on the Möbius band is represented by the point with rectangular coordinates (θ, t), the Möbius band, except for the line segment with longitude π(or $-\pi$), has a rectangle defined by $-\pi < \theta < \pi$, $-1 \leq t \leq 1$ as its map. As P moves across longitude π, $|t|$ varies continuously with P, but the positive and negative values of t are reversed. The rectangular map of the Möbius band will be complete if we add the left and right edges with the understanding that (π, t) and $(-\pi, -t)$ represent the same point on the band. An edge equation for the Möbius band is $abac^{-1} = 1$. Any point set in a Euclidean space that can be represented by a quadrilateral with this edge equation is topologically a Möbius band.

The identification of edges in the map of the Möbius band determines two distinct vertices, P and Q. Because the terminal vertex Q of b is the initial vertex of c and the terminal vertex P of c is the initial vertex of b, the edges b and c together form the single boundary curve of the Möbius band.

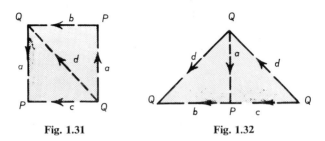

<table>
<tr><td>Fig. 1.31</td><td>Fig. 1.32</td></tr>
</table>

This boundary curve may be represented as a single edge if we cut the quadrilateral into two triangles and then reassemble the triangles in a different way. We cut the quadrilateral along the diagonal d and paste the triangles together so that the two occurrences of a coincide. Because b and c are not matched with other edges, the pattern of edge and vertex identification would be unchanged if b and c were merged into a single edge e, with P no longer a vertex. Thus the Möbius band has the alternate form of a triangle with the edge equation $dde = 1$. In this equation the unmatched edge e, which starts and finishes at Q, is the only boundary curve.

Although the cylinder and Möbius band are both represented as quadrilaterals with one pair of opposite edges identified, we have found a topological difference between them: the cylinder has two boundary curves, whereas the Möbius band has only one. Because the equations of the sphere and torus had no unmatched edges, neither the sphere nor the torus has any boundary curves. Thus the number of boundary curves is a topological criterion that distinguishes the cylinder and Möbius band from the sphere and torus. Surfaces such as a sphere and a torus with no boundary curves are called *closed surfaces*.

Visual comparison of paper models of a cylinder and a Möbius band is instructive. A cylinder may be made by bending a strip of paper around and pasting the ends together. A Möbius band may be made the same way except that a twist by π radians should be made in the strip before its ends are joined.

We return now to the Möbius band inscribed in a torus. If the points of this Möbius band are removed from the toroidal solid, the remaining points form a single solid. In Figure 1.30 the Möbius band divides the front portion of the solid into two parts, one nearer the origin than the other. As these parts are extended around to longitude π and beyond, the descriptions "nearer to" and "farther from" the origin are reversed and the two parts are fitted together as a single solid. Although we can paint two sides of a patch of the Möbius band two different colors, this painting pattern cannot be continued all the way around because the two colors, like the two parts of the toroidal solid, will merge together. The impossibility of painting the Möbius band inscribed in a torus with two colors that do not meet may be checked empirically with the help of crayons and a paper model of the band. "A. Botts and the Moebius Strip," by W. H. Upson, an anecdote based on this coloring property, was published in the *Saturday Evening Post* in 1945 and has been reprinted in *Fantasia Mathematica*, edited by Clifton Fadiman, Simon and Schuster, New York, 1958. The Möbius band inscribed in a Euclidean torus is an example of a *one-sided surface*.

Although the properties of one-sidedness and two-sidedness distinguish the Möbius band in a Euclidean torus from the cylinder in a Euclidean torus, we shall find that one-sidedness and two-sidedness are not topological properties of surfaces. In Chapter 7 we discuss a three-dimensional space that contains both one- and two-sided cylinders and one- and two-sided Möbius bands. Although this space, considered as a whole, is radically different from Euclidean three-space, each local portion resembles a portion of three-dimensional Euclidean space. One-sidedness and two-sidedness are embedding properties that depend not only on the surface but also on the space in which the surface is located and the way in which the surface is embedded in this space.

Although examples of one-sided and two-sided embeddings of a surface are not given until Chapter 7, we can illustrate the principle by examples of one-sided and two-sided curves on a surface. On the cylinder inscribed in the torus the axial circle of the torus divides the cylinder into two surfaces, one consisting of points with $z > 0$ and the other of points with $z < 0$. In contrast, the axial circle of the torus does not divide the inscribed Möbius band into two surfaces. The most convincing demonstration is given by cutting down the middle of a paper model of a Möbius band—the band will *not* fall apart. One-sidedness or two-sidedness is not a topological property of the simple closed curve, for we have found that the axial circle of the torus is two-sided as a curve on the inscribed cylinder and is one-sided as a curve on the inscribed Möbius band.

A simple closed curve on a surface is one-sided if a narrow strip on the surface with the curve down the middle is a Möbius band. If such a strip is a cylinder, the curve is two-sided. The role the narrow strip plays in the definition is shown by the case of a meridian on a torus. If the meridian is removed from the torus, only a single surface remains; hence the meridian appears to be one-sided. If attention is limited to the strip between two nearby meridians, with the given meridian in the middle of the strip, the strip is a cylinder divided by the given meridian into two cylinders, one on each side of the meridian. Hence by our definition, a meridian on a torus is a two-sided curve.

On the Möbius band inscribed in the torus consider the curve C, which is the intersection of this Möbius band and the sphere, $(x - 2)^2 + y^2 + z^2 = \frac{1}{25}$. The strip of points on the Möbius band with a distance no greater than $\frac{1}{10}$ from this curve is approximately a plane annulus. Hence the strip is topologically a cylinder and the curve is two-sided (Figure 1.33).

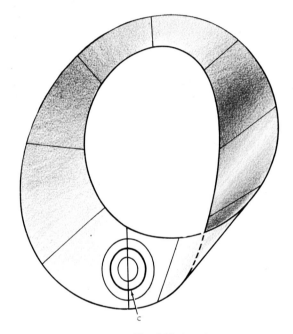

Fig. 1.33

On the other hand the axial circle of the torus is one-sided as a curve on the Möbius band. Thus the Möbius band has both one-sided and two-sided curves. It can be shown,† however, that all simple closed curves on the cylinder, sphere, or torus are two-sided.

Because a closed curve down the middle of a Möbius band does not divide the Möbius band into two surfaces, an observer traveling along

---

† With difficulty. See the sections on the Jordan curve theorem and the Schoenflies theorem in G. T. Whyburn, *Topological Analysis*, Princeton University Press, Princeton, New Jersey, 1958.

such a curve cannot consistently separate the Möbius band into points to his left and points to his right as he travels the curve. If he tried, he would find after one trip around that he would be reversing his previous definition of left and right. The Möbius band is the simplest example of a *nonorientable surface*. More generally, a surface is *nonorientable* if it contains a strip that is topologically a Möbius band. Another statement of this definition is that a surface is nonorientable if it contains a one-sided simple closed curve. *Orientable surfaces* are surfaces that are not nonorientable. Since a strip of points near a simple closed curve on an orientable surface is topologically a cylinder, the curve divides the strip into two surfaces, one of which may be defined as being to the left of the curve and the other to the right.

When the definition of nonorientability is extended in Chapter 7 to apply to three-dimensional spaces, we shall find that a necessary and sufficient condition for a three-dimensional space to be nonorientable is that it contains one-sided cylinders and two-sided Möbius bands.

1.5   Additional Representations of the Möbius Band

The cylinder and Möbius band, like the sphere and torus, can be divided into polygons. For example, Figures 1.34 and 1.35 show rectangular maps

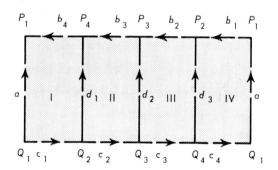

Fig. 1.34 Rectangular map of a cylinder

of a cylinder and Möbius band divided into quadrilaterals and triangles, respectively. The cylinder and Möbius band have Euler characteristics that equal $N_0 - N_1 + N_2$ for every subdivision of the surface into polygons. These special subdivisions have $N_0 = 8$, $N_1 = 12$, $N_2 = 4$ for the cylinder and $N_0 = 7$, $N_1 = 17$, $N_2 = 10$ for the Möbius band. Both the cylinder and the Möbius band therefore have Euler characteristic zero.

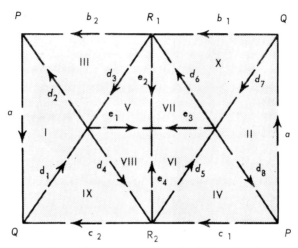

Fig. 1.35 Rectangular map of a Möbius band

If the rectangles of Figure 1.34 are squares of side one, the rectangular map can be bent along the edges d_1, d_2, and d_3 so that the adjacent squares will be perpendicular and the two edges labeled a will coincide. The result is a topological cylinder consisting of the four vertical faces of a cube (Figure 1.36).

Four isosceles right triangles, I, II, III, and IV (with hypotenuse of length one unit), can be fitted together with six equilateral triangles, V, VI, VII, VIII, IX, X (with side of length one), according to the pattern in Figure 1.35, to form the three-dimensional model of the Möbius band shown in Figure 1.37.†

To visualize this representation of the Möbius band we should not rely on the illustration but should construct a model of cardboard or better still of a rigid, transparent plastic. Note that the boundary curve is the perimeter of the Euclidean triangle with vertices Q, R_1, and R_2. Because we shall want to attach a Möbius band to other surfaces along its boundary curve, this band with its boundary curve flattened out into a plane is apparently ideal for drawing figures. Inspection of a cardboard model reveals a major difficulty. Whenever we try to paste the boundary curve of the Möbius band onto a boundary curve of another surface, such as a disk, the rest of the band gets in the way. The reason is that the boundary curve and the remainder of the Möbius band are linked. Instead of

---

† For a further explanation of this model of a Möbius band made of plane triangles see B. Tuckerman, *The American Mathematical Monthly*, **55** (1948), 309–311.

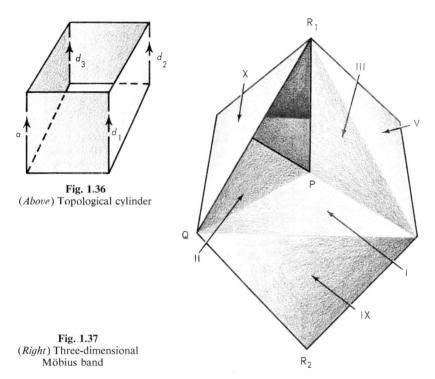

Fig. 1.36
(*Above*) Topological cylinder

Fig. 1.37
(*Right*) Three-dimensional
Möbius band

formally defining the term "linked," we shall describe an experiment that will demonstrate the meaning. Draw lines on a long rectangular strip of paper to divide it into three parallel strips, each one third as wide as the original strip. The outer pair of strips represents the edges of the middle strip. After pasting the ends of the original strip together to form a Möbius band, carefully cut the band along the lines you have drawn. Even after the edge and the rest of the Möbius band have been cut apart they cannot be separated because they are linked.

To find a model of the Möbius band suitable for connection with other surfaces, we have two alternatives. Either we construct an accurate model in four-dimensional Euclidean space or a slightly fudged model in three-dimensional space. Electing the second option, we start by dividing the rectangular map of the Möbius band into four quadrilaterals, as in Figure 1.38. Next we reassemble the four quadrilaterals as in Figure 1.39 so that the two occurrences of a_2 coincide. We bend this rectangle around as the north temperate zone of a sphere until the two edges (Figure 1.40) labeled a_1 coincide. Next, the points labeled S are moved up meridians of the

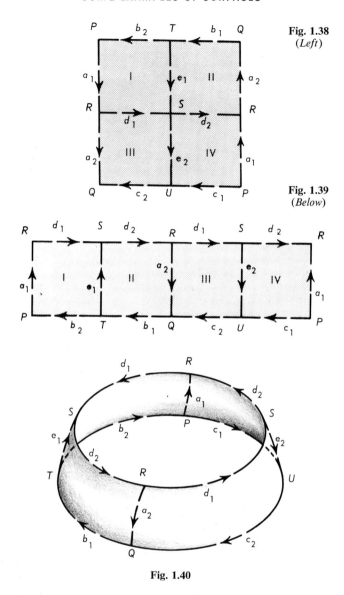

Fig. 1.38
(*Left*)

Fig. 1.39
(*Below*)

Fig. 1.40

sphere toward the north pole (Figure 1.41). Either the pair of edges labeled d_1 or the pair labeled d_2 could be brought together, but if one pair is pasted together the matching of the other pair is blocked. We fudge the model by allowing all four of these edges to coincide along the upper

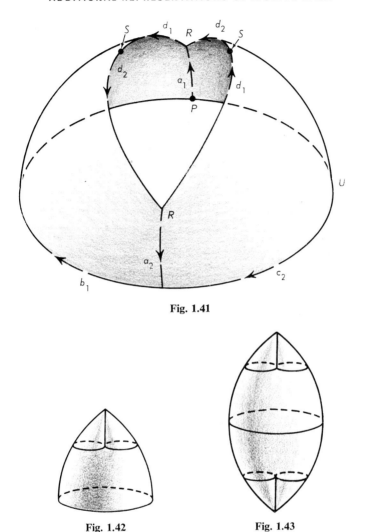

Fig. 1.41

Fig. 1.42   **Fig. 1.43**

quarter of the polar axis, as in Figure 1.42. On this model we must understand that the vertical edge actually represents two edges on the Möbius
band and that a point crossing the edge from the front of the surface on
the left or right passes to the back on the opposite side. This representation
of the Möbius band is an improper surface which crosses itself. Because
the surface intersects itself and resembles a bishop's hat, it was named a
crosscap. We use Möbius band and crosscap as synonyms. The boundary

curve is a Euclidean circle with the rest of the crosscap above this circle. Hence this crosscap can easily be attached to another surface along a boundary curve.

One simple example of a closed nonorientable surface may be obtained by fitting two crosscaps together along their boundary curves. This surface is named the *Klein bottle* (Figure 1.43) after its inventer, the German mathematician Felix Klein (1849–1925). Using the edge equations $d_1 d_1 e_1 = 1$ and $d_2 d_2 e_2 = 1$ for two crosscaps, we can signify that they are

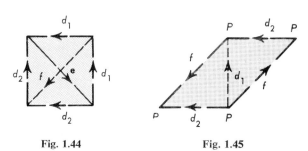

Fig. 1.44   Fig. 1.45

fitted together by replacing e_1 and e_2 with a single symbol e. Figure 1.44 is a rectangular representation of the Klein bottle with equation $d_1 d_1 d_2^{-1} d_2^{-1} = 1$. Delete the diagonal e, cut the rectangle along the diagonal f, and fit the triangles together so that the edges labeled d_1 coincide (Figure 1.45). We find that the Klein bottle may be represented as a quadrilateral with the edge equation $f d_2 f d_2^{-1} = 1$. The same vertex must be the initial vertex of both edges labeled f. Hence the top right and lower left vertices are both P. Because the initial vertices of the two edges labeled d_2 are the same, the lower right vertex must be the same as the upper right vertex P. Similarly, the upper left vertex is also P. Because there is one vertex, a pair of edges (f and d_2), and one quadrilateral, the Klein bottle has Euler characteristic 0. The torus and Klein bottle may be formed by identifying opposite edges of a quadrilateral and both have the same Euler characteristic.

1.6   The Projective Plane

The edge equations $aba^{-1}b^{-1} = 1$ and $abab^{-1} = 1$ specify the torus and Klein bottle as quadrilaterals with opposite edges identified. There remains a third way of matching the pairs of opposite edges. The surface with the equation $abab = 1$ is called the projective plane (Figure 1.46). In this form the projective plane has two vertices, two edges, and one quadrilateral. Because its Euler characteristic is 1, the projective plane differs topologically

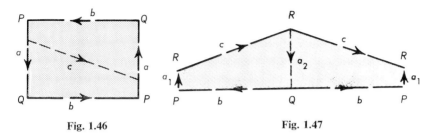

Fig. 1.46   Fig. 1.47

from the sphere, torus, cylinder, Möbius band, and Klein bottle. Draw the line segment c to join a pair of points, one on the left edge and one on the right, which represents the same point of the projective plane. Because c corresponds to a one-sided closed curve on the projective plane, the projective plane is nonorientable. The absence of unmatched letters in the edge equation shows that the projective plane is a closed surface.

Let the vertex R, where c meets a, divide a into the two edges a_1 and a_2. Cut the rectangle along the edge c and paste the pieces together so that the identification of edge a_2 is realized (Figure 1.47). Stretch this polygon around the south temperate zone of a Euclidean sphere so that the two edges labeled c form the equator, the two edges labeled b form the Tropic of Capricorn, the edge a_2 is an arc of the 0-meridian, and the two edges labeled a_1 coincide as an arc of the π-meridian (Figure 1.48). The two edges labeled b may be brought together as if they were the edges of two pages of a book with binding along the chord PQ of the sphere (Figure 1.49). We now erase the edges a_1, a_2, b_1, and the vertices P and Q from the subdivision of the projective plane (Figure 1.50). The model we have found for the projective plane is a southern hemisphere, with antipodal points on the equator identified. (A pair of diametrically opposite points on a Euclidean circle or sphere is called a pair of *antipodal points*). Hence a two-sided polygon with edge equation $cc = 1$ is a projective plane (Figure 1.51).

As the Euclidean hemisphere used in representing the projective plane, select the southern hemisphere of $x^2 + y^2 + z^2 = 1$. Consider the plane $z = -1$ tangent to this hemisphere at the south pole. A nonhorizontal line through the center O of the hemisphere intersects the hemisphere at a point P and the plane at a point P'. The point P' is the *gnomonic image* of P and the correspondence of P to P' is the *gnomonic projection* of the hemisphere onto the plane. Because a nonhorizontal plane of lines through O intersects the hemisphere in half a great circle and the plane in a line, the gnomonic images of the great circles on the hemisphere are the straight

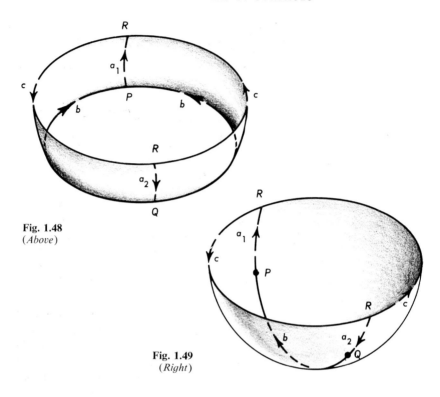

Fig. 1.48
(*Above*)

Fig. 1.49
(*Right*)

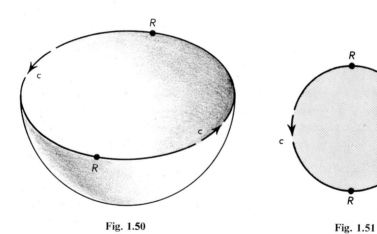

Fig. 1.50   Fig. 1.51

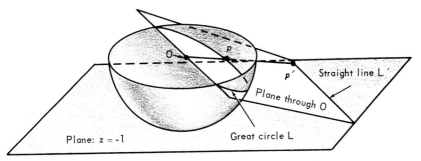

Fig. 1.52

lines in the plane (Figure 1.52). Now the shortest curve, or geodesic, on a sphere joining two points is an arc of a great circle, whereas in the plane the geodesic between two points is the straight-line segment joining them. A gnomonic map of a hemisphere of the earth has the useful property of representing geodesics on the earth by straight lines on the map. Figure 1.53 is a gnomonic map of the portion of the earth south of latitude $-\pi/8$. A gnomonic map is not conformal, and shapes on the earth are distorted.

Because the hemisphere with antipodal points on the equator identified is a projective plane, gnomonic projection maps all points of the projective plane except the equatorial points into the Euclidean plane $z = -1$. To make the representation of the projective plane complete we must add to the Euclidean plane new points that will be gnomonic images of the equatorial points. We now wish to find the properties of these new points.

A horizontal diameter D of the hemisphere meets the equator in two antipodal points P_1 and P_2 that represent a single projective point P. The great circles through P_1 and P_2 are the intersections of the hemisphere with planes through line D. This family of planes has the horizontal diameter D in common and intersects the horizontal plane $z = -1$ in the family of lines of $z = -1$, which are parallel to D (Figure 1.54). Thus point P determines a family of parallel lines in the Euclidean plane. Because the equatorial points representing P are common to the family of great circles corresponding to the family of parallel lines, the gnomonic image of P should be a point common to the family of parallel lines. The points added to the Euclidean plane, one for each family of parallel lines, are called *points at infinity*. This set of points is the image of the equator, and points at infinity are said to form the *line at infinity*. The geometry of the Euclidean plane with the line at infinity adjoined has the property that two distinct lines always determine a unique point and two distinct points always determine a unique line. In this geometry, called *projective plane*

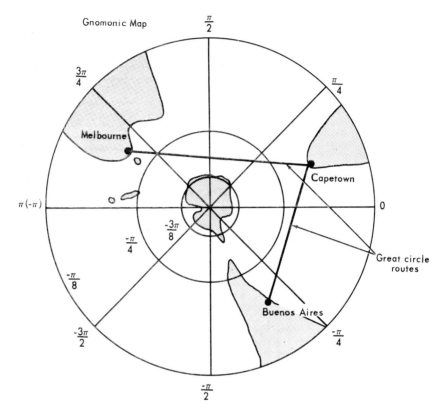

Fig. 1.53
Gnomonic map

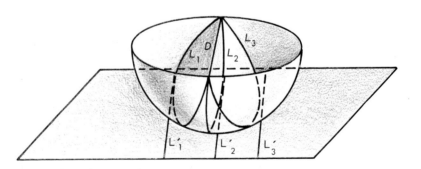

Fig. 1.54

geometry, many theorems of Euclidean plane geometry are simplified, for there is no need to consider separately cases in which lines of the theorem are intersecting or parallel.

We have arrived at the projective plane as a Euclidean plane with a line at infinity adjoined from the wrong historical direction. Projective geometry was developed by Desargues (1593–1662), Pascal (1623–1662), and Poncelet (1788–1867). In 1822 Poncelet introduced the projective plane as the extension of the Euclidean plane appropriate for the study of projective geometry.† The first study of topology as an organized discipline was made by Poincaré in the 1890's.

Starting from the Euclidean plane, we created topologically different surfaces by adding points at infinity in different ways. When a single point at infinity was added, the result was a sphere. If there is a line at infinity with one point at infinity on each family of parallel lines of the Euclidean plane, the surface is a projective plane.

We now return to Figure 1.47, erase the edge a_2, and draw the horizontal edge d to join the tops of the two edges labeled a_1. By the procedure used before the bottom polygon with equation $da_1^{-1}bb^{-1}a_1 = 1$ can be

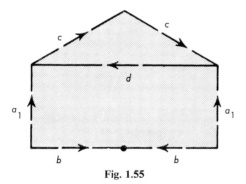

Fig. 1.55

changed into a southern hemisphere with d as the equator. The top polygon with equation $ccd = 1$ is a crosscap. Thus a projective plane is formed if the boundary curve of a crosscap is fitted onto the equator of a hemisphere.

In Section 1.3 we showed that six colors are sufficient to color the countries of any map on a sphere. Because the only property of the sphere used in the proof was the positivity of the Euler characteristic, the proof is equally valid for the projective plane. Figure 1.56 shows a map on the

† For a brief account of Poncelet's life and his contribution to projective geometry see Chapter 13 of *Men of Mathematics*, E. T. Bell, Simon and Schuster, 1937.

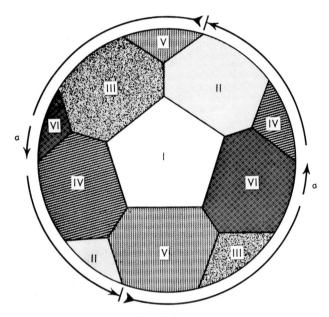

Fig. 1.56
Map on projective plane

projective plane which cannot be colored with fewer than six colors. The map has six pentagonal countries, each pair of which shares a common frontier.

This map is also an example of a regular subdivision of the projective plane into 6 pentagons, 15 edges, and 10 vertices, each of order 3. For a regular subdivision of the projective plane into polygons with k edges and vertices of order j the integers N_0, N_1, N_2 must satisfy the equations

$$2N_1 = kN_2, \qquad 2N_1 = jN_0, \qquad N_0 - N_1 + N_2 = 1.$$

Because the corresponding equations for the sphere were identical, except that $N_0 - N_1 + N_2 = 2$, a solution N_0, N_1, N_2 related to the sphere may be obtained by doubling a solution for the projective plane. A solution for the sphere with N_0, N_1, N_2 even integers may be converted into a solution for the projective plane by dividing these even integers by two. A check would show that a regular subdivision of the projective plane corresponds to each arithmetic solution for N_0, N_1, N_2. Figures 1.57, 1.58, and 1.59 show three such subdivisions.

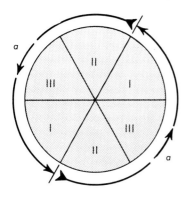

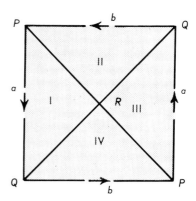

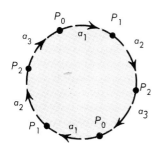

Fig. 1.57
(*Above, left*) $N_0 = 1$, $N_1 = N_2$, $j = 2N_1$, $k = 2$

Fig. 1.58
(*Left*) $N_0 = N_1$, $N_2 = 1$, $j = 2$, $k = 2N_2$

Fig. 1.59
(*Above, right*) $N_0 = 3$, $N_1 = 6$, $N_2 = 4$

EXERCISES

Section 1.1

1. Describe the locus of the equation

$$z^2 = [(x - 2)^2 + y^2 - 1][(x + 2)^2 + y^2 - 1](16 - x^2 - y^2).$$

2. A great circle on a Euclidean sphere is the intersection of the sphere with a plane through the origin. Show that the angle at which the great circle crosses the equator equals the maximum of the latitude of points on the great circle.

3. On the map of a sphere $x^2 + y^2 + z^2 = 1$, with latitude and longitude as rectangular coordinates, draw the image of a great circle on which $z = x$.

4. Draw the same great circle on the map with latitude and pseudolongitude as rectangular coordinates.

5. Consider the cross sections of the torus

$$z^2 = (\sqrt{x^2 + y^2} - 1)(3 - \sqrt{x^2 + y^2})$$

in the planes $ax + bz = 0$. If $|a/b| < \tan(\pi/6) = 1/\sqrt{3}$, the cross section is a pair of closed curves with one inside the other. If $|b/a| < \sqrt{3}$, the cross section is a pair of closed curves, neither of which is inside the other. Sketch the cross section in the intermediate case in which $b = \sqrt{3}a$.

6. A *rhumb line* on a Euclidean sphere is a curve that crosses each meridian at the same angle. A navigator follows a rhumb line by always traveling in a fixed compass direction.
   a. On a stereographic map of the southern hemisphere draw a rhumb line which starts from the equator in a southwesterly direction.
   b. Show that this rhumb line from the equator to the south pole has length $(\pi/2)\sqrt{2}R$, where R is the radius of the sphere.
   c. Which rhumb lines on a sphere are circles?

7. Derive Formula 1 when P is a point in the southern hemisphere.

Section 1.2

1. A surface S is formed from the four triangles 123, 234, 341, 412. (123 denotes the triangle with vertices labeled 1, 2, 3. Two triangles with a pair of vertices in common intersect in the edge joining these vertices.) What is the topological nature of S?

2. Consider two Euclidean loci, one defined by the parametric equations

$$x = \sin u, \qquad y = (\sin v)(2 + \cos u), \qquad z = (\cos v)(2 + \cos u)$$

where

$$-\pi \le u \le \pi \quad \text{and} \quad -\pi \le v \le \pi,$$

and the second by

$$x = \sin u, \qquad y = \sin v, \qquad z = u^2 + v^2$$

where

$$-\pi \le u \le \pi \quad \text{and} \quad -\pi \le v \le \pi.$$

One of these loci is topologically a torus and one is not. Which one is the torus? Explain why the other is not a torus.

3. What is the topological nature of the Euclidean locus with the parametric equations

$$x = u^2 + v^2, \qquad y = (u^2 + v^2 - 1)u, \qquad z = v,$$

where $u^2 + v^2 \leq 1$?

Section 1.3

1. Give an example of a map on the sphere that cannot be colored with less than four colors.

2. Let n great circles be drawn on a Euclidean sphere so that no three are concurrent. Into how many polygons do these circles divide the sphere? HINT: Every pair of great circles intersects in a pair of points. Also, the Euler characteristic of the sphere is 2.

3. Fit 32 cardboard squares together in the pattern of Figure 1.60 to make a model of the torus. Do these squares form a regular subdivision of the torus?

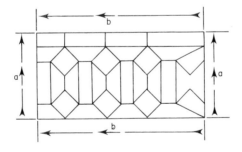

Fig. 1.60

4. Show that the order of the vertices must be at least 6 in a regular triangular sub-division of a surface with negative Euler characteristic.

5. A *triangulation* of a surface S is a subdivision of S into triangles so that each edge has two distinct vertices, no two distinct edges have the same pair of vertices, and no two distinct triangles have the same triple of vertices.

(a) In Figure 1.61 a torus represented as a hexagon with edge equation $abca^{-1}b^{-1}c^{-1} = 1$ has been triangulated into 14 triangles. Show that this is a minimal triangulation in the sense that the torus has no triangulation with fewer than 14 triangles.

HINT: For any triangulation of the torus

$$N_0 - N_1 + N_2 = 0, \qquad 2N_1 = 3N_2, \qquad N_1 \leq \frac{N_0(N_0 - 1)}{2}.$$

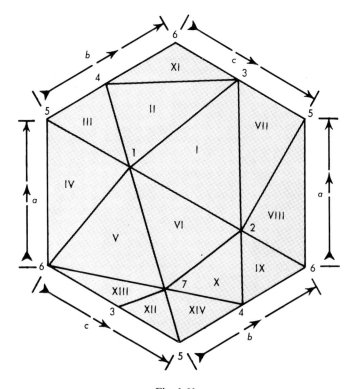

Fig. 1.61

(b) Find a minimal triangulation of the sphere.
(c) A triangulation Δ of a surface S consists of the triangles 123, 234, 345, 451, 512, 136, 246, 356, 416, and 526. (123 denotes a triangle with vertices labeled 1, 2, 3.) Show that S is neither a sphere nor a torus. Prove that Δ is a minimal triangulation of S.

Section 1.4

1. Draw a map with five countries on a Möbius band such that the map cannot be colored with fewer than five colors.

2. Two Euclidean loci are defined by the following parametric equations:

(a) $$x = \cos u, \qquad y = uv, \qquad z = (u^2 - \pi^2)(v + 2),$$

where

$$-\pi \le u \le \pi \quad \text{and} \quad -1 \le v \le 1.$$

(b) $$x = \sin u, \qquad y = |u(v + 3)| \qquad z = v,$$

where

$$-\pi \le u \le \pi \quad \text{and} \quad -1 \le v \le 1.$$

Which locus is a cylinder and which is a Möbius band?

3. (a) Why is the Euclidean locus described by the following parametric equations not a Möbius band?

$$x = \sin u, \qquad y = uv, \qquad z = |u|,$$

where

$$-\pi \le u \le \pi \quad \text{and} \quad -1 \le v \le 1.$$

(b) With the help of a stapler, convert a paper Möbius band into a topological model of this locus. What topological effect has this change on the boundary curve of the Möbius band?

Section 1.5

1. On a long strip of paper draw two parallel lines to divide the strip into three strips, each one third as wide as the original. After making one full twist (by 2π radians) in the strip, paste the two ends together. Cut the model along the lines on the paper. How many pieces result? How are they linked together? How many boundary curves did the surface have before they were cut off? What was the topological nature of the surface that was cut? Is the linking of a surface with its boundary curves a topological property of the surface?

2. In the text an experiment was described to show that the Möbius band inscribed in a torus is linked with its boundary curve. What is the topological nature of the surface that represented the boundary curve in this experiment?

3. Δ_1: 123, 234, 345, 451, 512,

Δ_2: 123, 234, 345, 456, 561, 612.

(a) Which of Δ_1 and Δ_2 triangulates the cylinder and which triangulates the Möbius band?

(b) In each triangulation list in sequence the edges of each boundary curve of the surface.

(c) Prove that Δ_1 and Δ_2 are minimal triangulations.

HINT: Let N_1' be the number of interior edges and N_1'' the number of boundary edges of any triangulation. Note that $N_0 - N_1 + N_2 = 0$, $3N_2 = 2N_1' + N_1''$, $N_1 \leq \frac{1}{2} N_0(N_0 - 1)$, $N_0 \geq N_1''$, and $N_1'' \geq 3$ on the Möbius band, $N_1'' \geq 6$ on the cylinder.

Section 1.6

1. Show that the following parametric equations describe a locus that is a projective plane in four-dimensional Euclidean space:

$$x_1 = u^2, \quad x_2 = uv, \quad x_3 = v^2 + u(u^2 + v^2 - 1), \quad x_4 = v(u^2 + v^2 - 1).$$

where $u^2 + v^2 \leq 1$.

2. Why does no parallel of latitude, except the equator, have a line as gnomonic image?

3. Represent the projective plane as the southern hemisphere of a Euclidean sphere, with antipodal points on the equator identified. Cut this model along a great circle from the equator to the south pole and back to the equator. How many pieces are there after the cutting is finished?

4. In projective geometry every pair of lines intersects in one and only one point. Let n lines be drawn in the projective plane so that no three are concurrent. Into how many polygons do these lines divide the projective plane?

5. Δ: 123, 234, 345, 451, 512, 136, 246, 356, 416, 526.

(a) Show that Δ is a triangulation of the projective plane.

(b) Show that Δ is a minimal triangulation.

2 THE CLASSIFICATION OF SURFACES

2.1 Surfaces and Their Equations

In Chapter 1 we studied many examples of surfaces but never really said exactly what a surface is. Because we define a surface as a set of points that can be "suitably" divided into polygons, we start by discussing polygons.

As a prototype of polygons with n sides, we adopt the disk D in the xy-plane defined by $x^2 + y^2 \leq 1$, together with n evenly spaced vertices $P_1 = (1, 0)$, P_2, P_3, ..., P_n which divide the circumference of D into edges $a_1, a_2, \ldots, a_n$. These polygons (Figure 2.1) with curved edges are selected as standard polygons in preference to those with straight edges because the polygons with curved edges include some with only one or two edges.

Definition. A homeomorphism f of the disk D onto a set π in a Euclidean space defines an n-sided *topological polygon* (π, f). The points $Q_i = f(P_i)$, $i = 1, 2, \ldots, n$, are the *vertices* of (π, f). The arcs $b_i = f(a_i)$, which are the images of the edges a_i of D, are the *edges* of (π, f). The *orientation* of b_i is

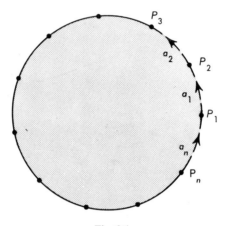

Fig. 2.1

the orientation induced by that on a_i. The equation $b_1 b_2 \cdots b_n = 1$ is an *edge equation* for (π, f).

Surfaces are formed by fitting polygons together along common edges. The way the polygons meet may be specified by listing the equations of the polygons, taking care to use the same letter to denote two occurrences of an arc as an edge of two different polygons. An exponent -1 is used to distinguish the opposite orientations of an edge. Although an edge may appear more than once in the equations of our surfaces, our definition of topological polygons does not permit an edge to occur more than once in the equation of a single polygon. Furthermore, no point may appear more than once as a vertex of a single polygon. With these restrictions, the

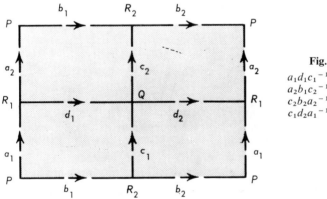

Fig. 2.2

$$a_1 d_1 c_1{}^{-1} b_1{}^{-1} = 1$$
$$a_2 b_1 c_2{}^{-1} d_1{}^{-1} = 1$$
$$c_2 b_2 a_2{}^{-1} d_2{}^{-1} = 1$$
$$c_1 d_2 a_1{}^{-1} b_2{}^{-1} = 1$$

torus could not be divided into fewer than four polygons. Figure 2.2 shows how a torus can be decomposed into four quadrilaterals. If polygons are generalized to allow the same edge to appear more than once on a polygon, we shall find that every surface may be represented as a single polygon. In the first chapter the idea of polygon with identified edges was exploited to represent the sphere, torus, cylinder, Möbius band, Klein bottle, and

Table 2.1

Sphere	Torus	Cylinder
$abb^{-1}a^{-1} = 1$	$aba^{-1}b^{-1} = 1$	$aba^{-1}c = 1$
Möbius Band	Klein Bottle	Projective Plane
$abac^{-1} = 1$	$abab^{-1} = 1$	$abab = 1$

projective plane as quadrilaterals with the edge equations (Table 2.1). An additional surface, the disk, may be represented as a quadrilateral with the edge equation $abcd = 1$.

To modify the definition of a topological polygon (π, f) to permit edge identifications we must relax the condition that f be a homeomorphism to allow the one-oneness of f to be violated on the edges and vertices of D.

Definition. A continuous mapping f of the disk D onto a set π defines a *singular topological polygon* (π, f) if f satisfies the following conditions:

1. Every point in π is $f(P)$ for some point P in D.
2. If P is an interior point of D and Q is any other point in D, then $f(Q) \neq f(P)$.
3. If a_j is an edge of D, there are two possibilities:
   (a) For every point P of a_j that is not a vertex, there is no point $Q(\neq P)$ in D such that $f(Q) = f(P)$.
   (b) For every point P of a_j other than P_j or P_{j+1} there is a unique point $P'(\neq P)$ in D such that $f(P') = f(P)$. Furthermore, as P moves from P_j to P_{j+1}, P' moves along an edge $a_k (k \neq j)$ either from P_k to P_{k+1} or from P_{k+1} to P_k. (P_{n+1} is interpreted as P_1).

The generalization to singular polygons will not lead to new surfaces, for every singular polygon can be divided into polygons without identified edges. To verify this let M_j be the midpoint of the arc a_j and draw in D the radii c_j terminating at P_j and d_j terminating at $M_j, j = 1, 2, \ldots, n$. The arcs $f(c_k)$ and $f(d_k)$ $k = 1, 2, \ldots, n$, divide (π, f) into polygons without identified edges. From now on *polygon* will mean singular topological polygon.

Before defining a surface as a point set that can be suitably divided into polygons, we preview the meaning of "suitably" by examples of unsuitable subdivisions. Figure 2.3 shows one triangle piercing the interior of another; Figure 2.4 has two triangles intersecting in a segment that is part of an

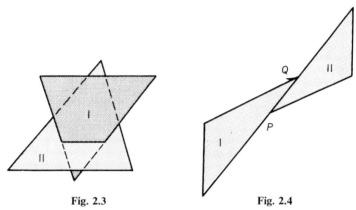

Fig. 2.3   **Fig. 2.4**

edge of each triangle. For a proper subdivision the intersection of any pair of polygons consists of a set (perhaps empty) of curves that are entire edges of both polygons and a set (perhaps empty) of additional points that are vertices of both polygons. This condition excludes the pairs of triangles in Figures 2.3 and 2.4 because the intersection of triangles I and II in Figure 2.3 contains interior points of the triangles and the intersection in Figure 2.4 is the segment PQ which is not a complete edge of either triangle. If the two triangles of Figure 2.4 are converted into quadrilaterals by adding P as a vertex of I and Q as a vertex of II, the pair of quadrilaterals is suitable. In Figure 2.5 a disk has been divided into one pentagon (I), one quadrilateral (II), and two triangles (III and IV). The pentagon and the quadrilateral intersect in the edge a (including its vertices) and the additional vertex P.

Figure 2.6 shows a sphere divided by the equator and four meridians into an octahedron with eight triangular faces. If the north and south poles are pushed together, the result is a *pinched sphere* divided into eight triangles (Figure 2.7). This subdivision is improper because one pair of vertices coincides, even though this coincidence is not required by the matching of edges of the various polygons. Another way of bringing the north and south poles together is to stretch the sphere parallel to the polar axis and then bend the sausage-shaped surface until the ends coincide (Figure 2.8). The result is a *strangled torus* formed by shrinking a meridian

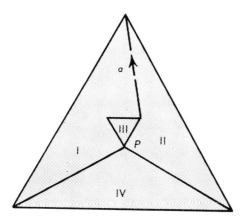

Fig. 2.5

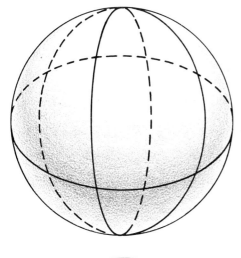

Fig. 2.6
(*Left*) Sphere divided by four
meridians and the equator

Fig. 2.7
(*Below, left*) Pinched sphere

Fig. 2.8
(*Below, right*) Strangled torus

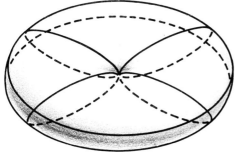

of the torus to a point. Figure 2.7 may be interpreted as a torus strangled by shrinking the π-parallel of latitude to a point. Topologically, there is no distinction between a pinched sphere and a strangled torus.

A third type of unsuitable subdivision is illustrated by a set of three or more rectangles bound together like the pages of a book (Figure 2.9). The

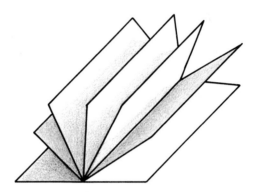

Fig. 2.9 Pages of a book

impropriety here is that a single curve is used more than twice as a polygonal edge.

Another exclusion is sets of polygons that form not one but two or more surfaces. Thus, if a tetrahedron and a cube do not intersect, the combined set of 10 polygons determines two surfaces. The separation into two parts may be verified formally by dividing the set of 10 polygons into a subset of six and a subset of four so that no polygon in the first subset shares an edge with a polygon in the second subset.

With this preview, we are now ready for the formal definition of a surface.

Definition. A set of points in a Euclidean space is a *surface* if the set can be subdivided into a finite number of polygons such that

1. Polygons intersect only in edges and vertices.
2. Polygons have common vertices only to the extent required by the common edges.
3. No curve is used more than twice as a polygonal edge.
4. The polygons cannot be divided into two sets of polygons with no edge in common.

Because Condition 2 prevents any subdivision of the pinched sphere from being suitable, the pinched sphere is not a surface. The pages (at least three) of a book are not a surface because Condition 3 cannot be satisfied. The same condition excludes the improper three-dimensional model of a crosscap shown in Figure 1.42. Although the two triangles of Figure 2.3 do not satisfy Condition 1, they can be subdivided into smaller polygons that do. Because Conditions 1 and 3 cannot be satisfied simultaneously, Figure 2.3 does not represent a surface.

The representation of polygons by edge equations leads to the representation of surfaces by systems of edge equations, one equation for each polygon in a subdivision of the surface. In the equations for a surface distinct edges must be represented by distinct letters and identified edges, by the same letter, except perhaps for an exponent -1, which indicates the direction of the edge on the perimeter of the polygon. Conditions 1 and 2 for the subdivision guarantee that the matching of edges determines completely how the polygons fit together. Condition 3 says that no letter occurs more than twice as an edge symbol. The letters appearing twice label interior edges, whereas those appearing only once stand for boundary edges. If there are no boundary edges, the surface is closed. Condition 4 means that no matter how the system of equations for a surface is divided into two sets, there is always at least one letter that appears in both sets of equations.

In writing an edge equation for a planar polygon, we must first select a starting point and then decide whether to travel clockwise or counterclockwise around the polygon. Because each selection of starting point and direction of travel determines an equation, many equations, in general, represent the same polygon. For example, the cylinder in Figures 2.10 and 2.11 has the equations

$$aba^{-1}c^{-1} = 1, \qquad ba^{-1}c^{-1}a = 1, \qquad a^{-1}c^{-1}ab = 1, \qquad c^{-1}aba^{-1} = 1,$$

$$cab^{-1}a^{-1} = 1, \qquad a^{-1}cab^{-1} = 1, \qquad b^{-1}a^{-1}ca = 1, \qquad ab^{-1}a^{-1}c = 1.$$

Each equation in the second line has the same starting point as the equation above it but lists the edges in the opposite direction. Algebraically, an equation in the second line is obtained by reversing the order of the letters in the equation above it and adding the exponent -1 if it does not appear and deleting it if it does. We refer to this algebraic process as inversion of the equation.

On a planar picture of a polygon we can travel from one vertex to another by two different paths along the perimeter, one clockwise and the other counterclockwise around the polygon. Additional equations for the

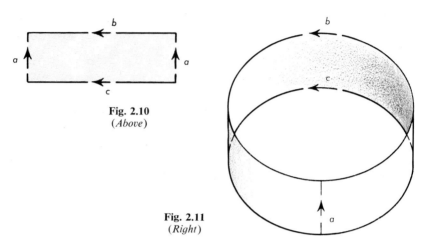

Fig. 2.10
(*Above*)

Fig. 2.11
(*Right*)

polygon are defined by equating each pair of edge sequences. The new equations for the cylinder of Figure 2.11 include

$$ab = ca, \qquad b = a^{-1}ca, \qquad ba^{-1} = a^{-1}c.$$

In the first equation the two edge sequences start at the lower right corner and end at the upper left. If the starting point is moved along the edge a to the upper right corner, the second equation is formed. The algebraic effect of the shift of starting point is to transpose the first letter a on the left side of the first equation to be the first letter a^{-1} of the right side of the second equation. If the finish of the two sequences is moved from the upper left along the edge a^{-1} to the lower left corner, the last letter a on the right side of the second equation is transposed to be the last letter a^{-1} on the left side of the third equation.

The original edge equations, in which a complete sequence is equated to 1, may be included among the new equations if we consider the complete sequence, which returns to its starting point, to be mated with the empty sequence (denoted by 1), which never leaves the starting point. In this broader context, a pair of transpositions changes the equation $aba^{-1}c^{-1} = 1$ first into $ba^{-1}c^{-1} = a^{-1}$ and then into $ba^{-1}c^{-1}a = 1$. Also four transpositions transform $aba^{-1}c^{-1} = 1$ first to $ba^{-1}c^{-1} = a^{-1}$, next to $a^{-1}c^{-1} = b^{-1}a^{-1}$, then to $c^{-1} = ab^{-1}a^{-1}$, and finally to $1 = cab^{-1}a^{-1}$ or $cab^{-1}a^{-1} = 1$. More generally any equation for a polygon can be converted by a sequence of transpositions into any other equation for the same polygon, provided only that the same letters denote the same edges in the two equations.

In a system of edge equations for a surface the particular letters selected to represent the edges have no significance. Any desired change of letters may be made by changing one letter at a time. When replacing one letter by another (perhaps with exponent -1), we must be sure that the new letter is not currently in use and that the old letter is replaced wherever it occurs. There is nothing to prevent a letter that has been replaced in one change from being reintroduced in a later change with a new geometric significance; for example, the c in the equation $aba^{-1}c^{-1} = 1$ for a cylinder is replaced by x^{-1}, to give $aba^{-1}x = 1$; then x is replaced by c, to give $aba^{-1}c = 1$. The c in the first equation and the c in the third equation stand for the same edge but with opposite directions. In the first change of letters $(x^{-1})^{-1}$ was written as x. This was proper, for two successive reversals of direction nullify each other.

The equations for a surface are more formally presented in the following definition:

Definition. A *combinatorial representation of a surface* is a system of edge equations such that (a) no letter appears more than twice; (b) if the system is divided into two sets of equations, there is at least one letter that appears in an equation of each set.

Two combinatorial representations are *trivially equivalent* if the first system of equations can be transformed into the second by a sequence of operations of the following types:

1. An initial (or terminal) letter on one side of an equation is transposed to appear in the opposite sense as the initial (or terminal) letter on the other side.

2. A new letter, not appearing in the equations, is substituted for an old letter wherever the old letter appears.

2.2 Combinatorial Equivalence

The equations of a surface may be transformed not only by changing the equations of a *given* polygonal subdivision but also by changing the subdivision itself. Examples of such changes were discussed in Chapter 1. We shall make these "cut-and-paste" operations more precise by defining combinatorial equivalence in terms of three pairs of inverse operations (see Figures 2.12–2.14).

The operation "cut a polygon in two" (Figure 2.12) changes the polygon and equation on the left into a pair of polygons and pair of equations on the right. The inverse operation, "paste two polygons together along a

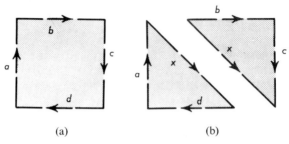

Fig. 2.12 (a) $abcd = 1$, (b) $axd = 1$, $x = bc$

common edge," replaces the polygons and equations on the right by the polygon and equation on the left. Algebraically, the cutting operation replaces the combination bc in the equation $abcd = 1$ by a single letter x and adds the new equation $x = bc$ to give the "value" of the new letter. The pasting operation eliminates one letter and one equation by substituting the "value" of x from the equation $x = bc$ into the equation $axd = 1$.

The operation "cut into a polygon" (Figure 2.13) transforms the poly-

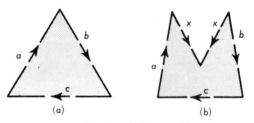

Fig. 2.13 (a) $abc = 1$, (b) $axx^{-1}bc = 1$

gon and equation on the left into the polygon and equation on the right. The inverse operation, "mend a cut," changes the polygon and equation on the right into the polygon and equation on the left. The cutting operation inserts the combination xx^{-1} into the equation $abc = 1$, whereas the x and x^{-1} adjacent to each other cancel each other in the pasting operation.

The operation "divide an edge" (Figure 2.14) replaces all occurrences (either one or two) of an edge b with a combination xy of new letters. Implicit in this change is the introduction of a new vertex as the terminal vertex of x and the initial vertex of y. Thus the triangles $abc = 1$ and $dbe = 1$ become the quadrilaterals $axyc = 1$ and $dxye = 1$. Because the second triangle and the second quadrilateral could have been represented by the equations $e^{-1}b^{-1}d^{-1} = 1$ and $e^{-1}y^{-1}x^{-1}d^{-1} = 1$, we see that substitution of $y^{-1}x^{-1}$ for b^{-1} is equivalent to substitution of xy for b. In the

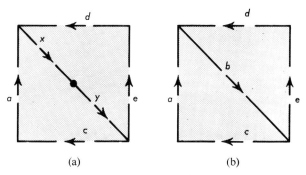

(a)   (b)

Fig. 2.14 (a) $axyc = 1$, $dxye = 1$, (b) $abc = 1$, $dbe = 1$

inverse operation, "suppress a vertex of order two," a pair of edges x and y, which always appear together either in the combination xy or the combination $y^{-1}x^{-1}$, is replaced by a single edge b, with b substituted for xy and b^{-1} substituted for $y^{-1}x^{-1}$. The vertex P between x and y is eliminated when the edges are united. The condition that x and y always appear as xy or $y^{-1}x^{-1}$ ensures that P has order 2; that is, P is on no edge other than x and y. In Figure 2.15 the vertex Q of order 3 could not be sup-

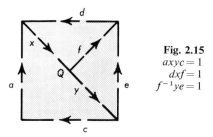

Fig. 2.15
$axyc = 1$
$dxf = 1$
$f^{-1}ye = 1$

pressed. The inapplicability of the suppression operation is shown algebraically by the separate appearances of x and y in the equations: $axyc = 1$, $dxf = 1$, and $f^{-1}ye = 1$.

Definition. Two combinatorial representations of surfaces are *combinatorially equivalent* if the equations of one representation can be transformed into the equations of the other by a sequence of trivial operations and combinatorial operations of the following six types (the algebraic versions of the cut-and-paste operations):

1. A new letter x is selected and an equation of the form $ABC = 1$ is replaced by two equations $AxC = 1$ and $x = B$. (Capital letters represent blocks of edge symbols.)

$1'$. A pair of equations $AxC = 1$ and $x = B$ is replaced by the equation $ABC = 1$.

2. A new letter x is selected and the combination xx^{-1} is inserted in an equation in any position.

$2'$. A combination xx^{-1} (or $x^{-1}x$) is deleted from an equation. The combination xx^{-1} standing alone on one side of an equation would be replaced by 1.

3. After two distinct new letters u and v are selected a letter x is replaced wherever it occurs by the combination uv. If x^{-1} occurs, it is replaced by $v^{-1}u^{-1}$.

$3'$. A selection is made of letters u and v which occur only in the combinations uv or $v^{-1}u^{-1}$. The combination uv is replaced by x whereas $v^{-1}u^{-1}$ is replaced by x^{-1}.

Definition. Two surfaces are *combinatorially equivalent* if they have combinatorial representations that are combinatorially equivalent.

A word of caution about our equations is necessary. The relation designated by the equal sign does not satisfy all the usual axioms of an equivalence relation. Although the symmetric law is satisfied, the reflexive law is generally false, and the transitive law cannot be applied to say that two expressions equal to 1 are equal to each other. Despite these drawbacks, the symbolic equations share enough of the usual properties of equations to make them convenient; for example, we may use transpositions to solve an equation for one of its letters x. By applying Operation $1'$ we may substitute this value of x in another equation, thereby eliminating one letter and one equation.

In Section 2.3 we show how the equations of a combinatorial representation may be reduced to a single canonical equation. Two surfaces will be combinatorially equivalent if and only if their canonical equations are trivially equivalent. Thus the canonical equations classify all surfaces.†

As a prelude to deriving the canonical equation of a surface, we look for features that might distinguish different surfaces. First, an equation such as $AaBaC = 1$, with a letter appearing twice in the same sense, shows that a Möbius band (or crosscap) can be cut from the surface. In Figure 2.16 the strip between the dotted lines is a Möbius band. If the equation could

---

† Our procedure for classifying surfaces is a simplification by A. W. Tucker of a method of Brahana (*Annals of Mathematics*, **23** (1921), pp. 144–168). E. F. Whittlesey has used an extension of the technique to classify two-dimensional complexes (*Mathematics Magazine*, **34** (1960), pp. 11–22, 67–80; *Proceedings of the American Mathematical Society*, vol. **9** (1958), pp. 841–845.

be rewritten in the form $A'aaB' = 1$, the crosscap (or Möbius band) $aad = 1$ (Figure 2.17) could be severed from the rest of the polygon with one cut rather than two. Lemma 1 in Section 2.3 establishes combinatorial equivalences that permit us to bring together the occurrences of a letter appearing twice in the same sense on the perimeter of a polygon.

When two occurrences of a letter a, once in each sense, can be brought together in an equation, the combination aa^{-1} can be deleted. Although a pair a and a^{-1} cannot always be brought together, Lemma 2 permits us to transform the equation so that no more than one letter is between a and a^{-1}. There are two types of combination in which inverse pairs are separated by one letter. Either two letters a and b, each occurring twice, appear

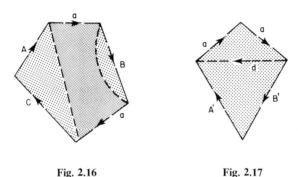

Fig. 2.16   Fig. 2.17

in an alternating combination $aba^{-1}b^{-1}$ or there is a boundary edge b between a and a^{-1} in a combination aba^{-1}.

Cutting a polygon with equation $Aaba^{-1}b^{-1}B = 1$ along the edge x in Figure 2.18 separates a polygon with equation $x = aba^{-1}b^{-1}$ or $aba^{-1}b^{-1}x^{-1} = 1$. In Figure 2.19(a) or 2.19(b) we recognize this as the equation of a torus with a disk bounded by x removed. A torus with a disk removed is called a *handle* when attached to another surface along the boundary curve x. When a combination $aba^{-1}b^{-1}$ can be assembled in an equation of a surface, the surface has a handle.

If a polygon of a surface with equation $Aaba^{-1}B = 1$, where b is a boundary edge, is cut in two as $AxB = 1$ and $x = aba^{-1}$, the cylinder $x = aba^{-1}$ is a cuff on the surface.

Geometrically interpreted, the reduction of the equations of a combinatorial representation of a surface to a single canonical equation is a systematic procedure for determining the number of crosscaps, handles, and cuffs on a surface.

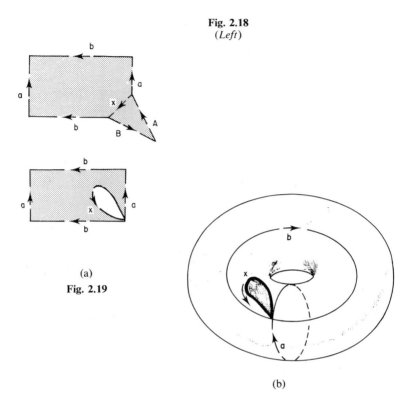

Fig. 2.18
(Left)

(a)

Fig. 2.19

(b)

2.3   The Canonical Equation

We start by proving two lemmas which state rules for moving blocks of letters adjacent to one occurrence of a letter x to a position adjacent to the other occurrence of x(as either x or x^{-1}).

Lemma 1. If a polygon of a surface has an equation of the form

$$AxBCxD = 1,$$

where A, B, C, and D are blocks (possibly empty) of edge symbols, the given surface is combinatorially equivalent to a surface with the same equations except that this special equation has been replaced by

$$AxCxB^{-1}D = 1,$$

where B^{-1} is the block of the inverses of the symbols in B written in the reverse order. Similarly, combinatorial equivalence allows an equation of the form

$$ABxCxD = 1,$$

to be replaced by

$$AxCB^{-1}x\,D = 1.$$

The first step is to use Operation 1 to replace the equation $AxBCxD = 1$ with the equations

$$AyCxD = 1$$

$$y = xB.$$

By transposing the symbols in B, one by one, we may rewrite the second equation as $x = yB^{-1}$. Operation 1′ replaces the two equations with

$$AyCyB^{-1}D = 1.$$

Because a single letter can occur at most twice in the equations for a surface, and we have replaced two occurrences of x, the letter x is no longer used. Therefore we can now replace y with x and rewrite the equation in the desired form:

$$AxBCxD = 1 \implies AxCxB^{-1}D = 1.$$

Figure 2.20 is proof that $ABxCxD = 1$ can be replaced by

$$AxCB^{-1}xD = 1.$$

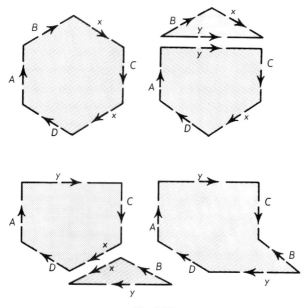

Fig. 2.20

In the final step of the proof the notation is changed by replacing y with x.

Lemma 2. Combinatorial equivalence permits an equation of the form

$$AxBCx^{-1}D = 1$$

to be replaced with the equation

$$AxCBx^{-1}D = 1.$$

Similarly, $ABxCx^{-1}D = 1$ may be replaced by $AxCx^{-1}BD = 1$.

By operation 1, $AxBCx^{-1}D = 1$ can be replaced by $AyCx^{-1}D = 1$ and $y = xB$. After the second equation is rewritten as $x^{-1} = By^{-1}$, Operation 1' replaces the two equations with the equation $AyCBy^{-1}D = 1$. Replacement of y with x gives the desired equation. A similar proof may be given for the second part of the lemma.

The two lemmas may be paraphrased as follows:

If a letter x occurs twice in the same sense on the boundary of a polygon, a block beside one occurrence of x may be moved to the same side of the other occurrence, provided the direction of the block is reversed.

If a letter x occurs once in each direction on the boundary of a polygon, a block may be moved from one side of the occurrence of x to the other side of the occurrence of x^{-1}. The direction of the block is not changed.

Lemma 3. Using combinatorial equivalence, we may represent any surface by a single equation or a single polygon with identified edges.

Consider any equation representing a polygon of a surface. The fourth condition in the definition of a surface ensures that some letter x in this equation will also appear in another equation. Eliminate the letter x between these two equations and reduce the number of equations by one. This can be repeated until only one equation remains.

We now show how Lemmas 1 and 2 may be used to reduce a single equation for a surface to a canonical form.

Step 1. Assemble the Crosscaps.

Suppose an equation for a surface has the form $ABsCsD = 1$. By Lemma 1 the block B may be moved from in front of the first s to give

$$AsCB^{-1}sD = 1.$$

Another application of Lemma 1 allows the block CB^{-1} to be transferred from behind the first s to give

$$AssBC^{-1}D = 1.$$

If $A = c_1 c_1 \cdots c_k c_k$, this combinatorial equivalence allows us to append to A a new pair of repeated letters $c_{k+1} c_{k+1}$. We can continue to enlarge the block A of pairs until the remainder of the equation contains no letter that is repeated twice in the same sense. An equation of the form

$$c_1 c_1 c_2 c_2 \cdots c_q c_q H = 1$$

may be replaced by the equations

$$y_1 = c_1 c_1, \qquad y_2 = c_2 c_2, \ldots, y_q = c_q c_q, \qquad y_1 y_2 \cdots y_q H = 1.$$

The equations $y_i = c_i c_i$ represent crosscaps that are attached to the rest of the surface along the edges y_i.

Step 2. Assemble the handles.

Suppose an equation for a surface has the form

$$ABsCtDs^{-1}Et^{-1}F = 1.$$

In this equation the letters s and t appear in both senses and the occurrences of s are separated by those of t. By Lemma 2 the block B may be moved from in front of s to behind s^{-1} to give

$$AsCtDs^{-1}BEt^{-1}F = 1.$$

This becomes

$$AsCtBEDs^{-1}t^{-1}F = 1$$

if Lemma 2 is used to move BE from in front of t^{-1} to behind t. Similarly, the block BED may follow s instead of preceding s^{-1}, thus giving

$$AsBEDCts^{-1}t^{-1}F = 1.$$

After a final move of the block $BEDC$ the equation becomes

$$Asts^{-1}t^{-1}BEDCF = 1.$$

We incorporate the block $sts^{-1}t^{-1}$ into A and repeat the procedure with a new pair of letters. If we start with the block $c_1 c_1 c_2 c_2 \cdots c_q c_q$ as A, Step 2 leads to the equation

$$c_1 c_1 c_2 c_2 \cdots c_q c_q a_1 b_1 a_1^{-1} b_1^{-1} a_2 b_2 a_2^{-1} b_2^{-1} \cdots a_p b_p a_p^{-1} b_p^{-1} H = 1,$$

where there is no pair of letters s and t in H in which occurrences of s and s^{-1} are separated by occurrences of t and t^{-1}. At the end of Step 1 no letters after c_q appeared twice in the same sense. Because Step 2 inverts no letters and does not change the initial block $c_1 c_1 \cdots c_q c_q$, no letter in H appears twice in the same sense.

The preceding equation may be replaced by

$$c_1 c_1 \dots c_q c_q x_1 x_2 \dots x_p H = 1,$$

$$x_1 = a_1 b_1 a_1^{-1} b_1^{-1}, \; x_2 = a_2 b_2 a_2^{-1} b_2^{-1}, \dots, x_p = a_p b_p a_p^{-1} b_p^{-1}.$$

The p equations $x_i = a_i b_i a_i^{-1} b_i^{-1}$ represent handles attached to the surface along the edges x_i.

Step 3. Using a crosscap as a catalyst, turn handles into crosscaps.

Suppose that an equation for a surface has been reduced by using Steps 1 and 2. If there is at least one crosscap and at least one handle, this equation has the form

Change ab to $\overline{ba}$
Collect $b\,b$
Collect $a\,a$

$$Accaba^{-1}b^{-1}G = 1.$$

By moving ab which is behind the second c to a position behind the first c, we derive

$$Acb^{-1}a^{-1}ca^{-1}b^{-1}G = 1.$$

The equation $Acac^{-1}ab^{-1}b^{-1}G = 1$ follows if $a^{-1}ca^{-1}$ before the last b^{-1} is placed in front of the first b^{-1}. When we shift c^{-1} before the second a to a place before the first a, the equation becomes

$$Accaab^{-1}b^{-1}G = 1.$$

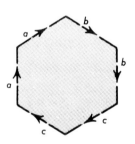

Fig. 2.21

We have changed the handle into two crosscaps by using the crosscap already present. By combining Steps 1, 2, and 3 we find that any surface equation can be written in one of two forms:

$$c_1 c_1 \dots c_q c_q H = 1,$$

$$a_1 b_1 a_1^{-1} b_1^{-1} \dots a_p b_p a_p^{-1} b_p^{-1} H = 1,$$

where neither crosscaps nor handles can be separated from H.

The convertibility of a handle, in the presence of a crosscap as a catalyst, into a pair of crosscaps may seem surprising. We digress to discuss an example geometrically. The hexagon in Figure 2.21 with equation $aabbcc = 1$ is a surface with three crosscaps. In Figure 2.22 a strip that crosses the edges a, b, and c has been shaded on the surface. The edges a, b, and c of Figure 2.21 have each been divided into three edges in Figure 2.22. The three parts of the strip have been assembled in Figure 2.23 to show that

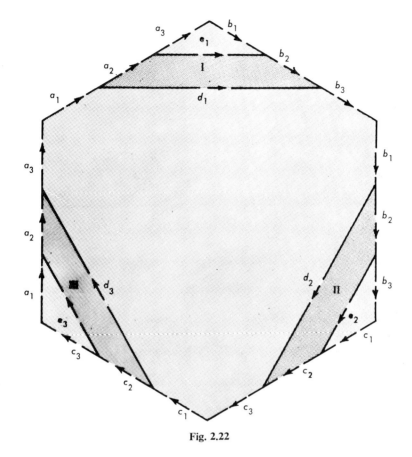

Fig. 2.22

the strip is a Möbius band. In Figure 2.24 the remaining four parts of the surface have been assembled as a single polygon with equation

$$a_3 c_3^{-1} e_3 d_1 b_3 a_3^{-1} e_1 d_2 c_3 b_3^{-1} e_2 d_3 = 1.$$

In Step 2 we found that an equation

$$ABsCtDs^{-1}Et^{-1}F = 1$$

is combinatorially equivalent to the equation

$$Asts^{-1}t^{-1}BEDCF = 1.$$

We apply this rule to the present equation with a_3 and b_3 as s and t, with empty blocks as A, B, and D, and with the blocks $c_3^{-1} e_3 d_1$, $e_1 d_2 c_3$, and

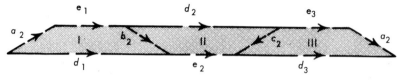

Fig. 2.23

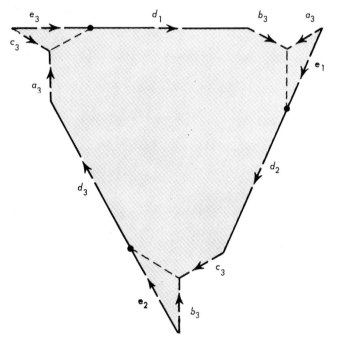

Fig. 2.24

e_2d_3 as C, E, and F, respectively. The result is the equation

$$a_3b_3a_3^{-1}b_3^{-1}e_1d_2c_3c_3^{-1}e_3d_1e_2d_3 = 1.$$

Deletion of $c_3c_3^{-1}$ gives

$$a_3b_3a_3^{-1}b_3^{-1}e_1d_2e_3d_1e_2d_3 = 1.$$

Repeated use of Operation 3' justifies substitution of a single letter f for
the block $e_1d_2e_3d_1e_2d_3$. The final equation

$$a_3b_3a_3^{-1}b_3^{-1}f = 1$$

represents a handle. We have found that the surface with three crosscaps is also a crosscap fitted to a handle.

Step 4. Assemble the cuffs.

After Steps 1, 2, and 3 have been performed, the remaining block H may still have letters that occur twice, once in each sense. Let d be a letter appearing twice with the fewest letters between d and d^{-1}. A letter t could not occur twice between d and d^{-1}, for there would be fewer letters between t and t^{-1} than between d and d^{-1}. A letter t between d and d^{-1} cannot appear outside the block from d to d^{-1}, for an outside occurrence of t^{-1} would give an additional handle. Because each of the letters between d and d^{-1} appears only once, the block between d and d^{-1} can be replaced by a single new letter e. The equation of the surface is now

$$ABded^{-1}C = 1,$$

where A is the block of crosscaps or handles. By Lemma 2 an equivalent equation is

$$Aded^{-1}BC = 1.$$

Because no letters are inverted in this equivalence, no additional crosscaps are created, and because the letters in the block BC are in the same order as before there are no additional handles. Repetition of this procedure yields an equation of the form

$$Ad_1e_1d_1^{-1} \dots d_re_rd_r^{-1}H = 1,$$

where no letter appears twice in H. If H is not an empty block, it may be replaced by a single new letter e. In this case let us append the block $d^{-1}d$ after e to obtain the equation

$$Ad_1e_1d_1^{-1} \dots d_re_rd_r^{-1}ed^{-1}d = 1.$$

By two transpositions the letter d may be moved from the last to the first position so that the equations start with one of the blocks

$$dc_1c_1 \quad \text{or} \quad da_1b_1a_1^{-1}b_1^{-1}.$$

In the first case Lemma 1 allows d in front of the first c_1 to be replaced by d^{-1} in front of the second c_1. A second application of this lemma justifies the substitution of d after the second c_1 for d^{-1} after the first c_1. In the second case Lemma 2 allows d to move first from in front of a_1 to between a_1^{-1} and b_1^{-1}, second, from in front of b_1^{-1} to between b_1 and a_1^{-1}, and third, from in front of b_1 to behind b_1^{-1}. In the two cases we see that the

$*$ *I think this needs 4 steps*

$$d\,a\,\bar{b}\,\bar{a}\,b \xrightarrow{} a\,b\,\bar{a}\,d\,\bar{b} \xrightarrow{\bar{b}} a\,b\,d\,\bar{a}\,\bar{b} \xrightarrow{\bar{a}} a\,d\,b\,\bar{a}\,\bar{b}$$

$$\xrightarrow{b} a\,b\,\bar{a}\,\bar{b}\,d.$$

letter d may be moved past a crosscap or a handle. Moving d past all the crosscaps or handles leads to the equation

$$Add_1e_1d_1^{-1}d_2e_2d_2^{-1} \cdots d_re_rd_r^{-1}ed^{-1} = 1.$$

Repeated use of Lemma 2 allows d to be moved past the blocks $d_1e_1d_1^{-1}$, $d_2e_2d_2^{-1}, \ldots, d_re_rd_r^{-1}$ to give the equation

$$Ad_1e_1d_1^{-1}d_2e_2d_2^{-1} \cdots d_re_rd_r^{-1}ded^{-1} = 1.$$

When the blocks $d_ie_id_i^{-1}$ are assembled, the remaining block can be empty or can contain a single letter. The second case need not be considered, for it can be converted to the first case by assembling one additional block ded^{-1}.

We have now shown that every surface may be represented by a canonical equation in one of two forms:

$$c_1c_1 \cdots c_qc_qd_1e_1d_1^{-1} \cdots d_re_rd_r^{-1} = 1$$

or

$$a_1b_1a_1^{-1}b_1^{-1} \cdots a_pb_pa_p^{-1}b_p^{-1}d_1e_1d_1^{-1} \cdots d_re_rd_r^{-1} = 1.$$

These equations could be replaced by the following systems of equations:

$$y_1 = c_1c_1, \ldots, y_q = c_qc_q, \qquad z_1 = d_1e_1d_1^{-1}, \ldots, z_r = d_re_rd_r^{-1},$$

$$y_1 \cdots y_qz_1 \cdots z_r = 1$$

or

$$x_1 = a_1b_1a_1^{-1}b_1^{-1}, \ldots, x_p = a_pb_pa_p^{-1}b_p^{-1},$$

$$z_1 = d_1e_1d_1^{-1}, \ldots, z_r = d_re_rd_r^{-1}, \qquad x_1 \cdots x_pz_1 \cdots z_r = 1.$$

The equations $y_i = c_ic_i$, $x_i = a_ib_ia_i^{-1}b_i^{-1}$, and $z_i = d_ie_id_i^{-1}$ represent crosscaps, handles, and cuffs attached to a sphere from which patches have been cut along the edges y_i, x_i, and z_i. The equation $y_1 \cdots y_qz_1 \cdots z_r = 1$ or $x_1 \cdots x_pz_1 \cdots z_r = 1$ represents the sphere from which the patches have been removed. We show this by combining the equation $y_1 \cdots y_qz_1 \cdots z_r = 1$ with the equations $y_1 = 1, \ldots, y_q = 1, z_1 = 1, \ldots, z_r = 1$ for patches. By combinatorial equivalence these equations can be reduced to the pair $y_1 = 1$ and $y_1 = 1$. These two identical equations represent two polygonal regions with a common boundary. If we think of the regions as northern and southern hemispheres both bounded by a common equator, we see that the surface is combinatorially equivalent to a sphere. Carrying the reduction process one step further, we have the equation $1 = 1$. We can think of it as an equation for the sphere. It arises as a canonical equation when $q = p = r = 0$. We have now proved Theorem 1.

Theorem 1. Every surface is combinatorially equivalent to a sphere with a finite number of patches replaced by crosscaps, handles, and cuffs. If the number of crosscaps is greater than zero, the handles can be replaced by pairs of crosscaps.

If there is at least one crosscap or Möbius band in the canonical equation, the surface contains a nonorientable piece, hence is nonorientable. Orientability is defined more formally in Section 2.4. If there are no cuffs on a surface, every edge occurs twice and the surface is closed.

2.4 Combinatorial Invariants of a Surface

In Section 2.3 we described a method of reducing the equations of a surface to a single canonical equation. To justify calling this equation "canonical" we must show that two different sequences of combinatorial equivalences cannot reduce the equations of a surface to different canonical equations. We prove it by showing that orientability, the number p or q called the *genus*, and r, the number of cuffs, are intrinsic properties of the surface which are invariant under combinatorial equivalence.

Definition. A surface is orientable if its equations can be changed by trivial equivalence so that each equation has 1 as its right-hand member and any letter x that appears twice occurs once as x and once as x^{-1}.

Under Operation 1 an equation $ABC = 1$ is replaced by the pair of equations $AxC = 1$ and $x = B$. We may rewrite this second equation as $Bx^{-1} = 1$. If, in the original equations for the surface, each letter appearing twice occurs once in each sense, the same property holds for the new set of equations, for the old letters appear in the same sense as before and the new letter occurs once as x and once as x^{-1}. On the other hand, suppose that $AxC = 1$ and $Dx^{-1}E = 1$ are two of the equations for a surface. By trivial equivalence the second equation may be changed to $EDx^{-1} = 1$ or $x = ED$. By Operation 1' the pair of equations is replaced by $AEDC = 1$. Because the sense of no letter has been reversed, a system of equations written in a form to show the orientability of the surface still demonstrates orientability after we use Operation 1' to eliminate one variable. Operations 2 and 2' cannot affect orientability for they simply introduce or eliminate a combination xx^{-1}. Because Operation 3 replaces each occurrence of a letter x with the combination uv taken in the same sense, orientability is invariant under Operation 3. Similarly, Operation 3' cannot change an orientable to a nonorientable surface. We can now conclude that an orientable surface stays orientable under combinatorial equivalence. If a nonorientable surface could become orientable by combinatorial equivalence, the reverse equivalence would make an orientable surface nonorientable. We now see that the orientability or nonorientability of a surface is combinatorially invariant. A surface is nonorientable only if it can be reduced to a canonical form with $q > 0$; that is, the surface contains a crosscap (or a Möbius band).

Before we can show that the number of cuffs and the number of handles or crosscaps are invariant under combinatorial equivalence, we must study the vertices of a surface. The definition of surface includes the condition that polygons have common vertices only to the extent required by the common edges. We must now study the implications of this condition.

Each edge on the boundary of a polygon has an initial and a terminal vertex. If an edge a appears in the opposite sense as a^{-1}, the initial and terminal vertices are reversed. In an equation for a polygon consecutive letters represent consecutive boundary edges that meet in a common vertex, namely, the terminal vertex of the first edge and the initial vertex of the second. This may be symbolized by inserting vertex symbols between adjacent edges in our equations. In an equation $A = 1$ the last and first edges are consecutive; therefore the same vertex symbol should be placed after the last letter and before the first. If the combinations ab and ac both appear in the equations for a surface, a vertex symbol P placed between a and b must also be inserted between a and c, for a has the same terminal point wherever it occurs. If the combination dc also appears, the vertex symbol P must be used again, for c has a unique initial vertex. If d appears a second time, this sequence of occurrences of P can be continued. Because only a finite number of positions for vertex symbols exists, the sequence must stop or start to repeat. The sequence can stop only by reaching a letter that occurs only once. The sequence repeats by reaching the original combination aPb. If d had been an edge appearing only once, we could have started with the combination dPc, extended the sequence of occurrences of P back through the combination aPb, and continued until we were again stopped by some singly appearing edge f which had P as initial or terminal vertex. The sequence must end and not repeat, for there are no other occurrences of d from which to return to the combination dPc. The procedure we have just described enables us to decide exactly which of the vertices of the polygons must coincide as a consequence of the edge identifications.

Example. Let a surface be represented by the equations

$$abcde = 1, \qquad ac^{-1}e^{-1} = 1.$$

If P is the vertex between a and b, P is also between a and c^{-1} and c and d. Because b and d appear only once, we have found all occurrences of P. If Q is the vertex terminating e and starting a in the first equation, Q also starts a and ends e^{-1} in the second equation. Hence Q starts e and finishes d in the first equation. The edge d appears only once, thus stopping our sequence in one direction. Because Q ends e in the first equation, we find that Q also belongs between c^{-1} and e^{-1} in the second. This implies that Q is between b and c in the first equation. Because b appears only once, we have now found the second end of the sequence of occurrences of Q. Using both edge and vertex symbols, we rewrite the equations in expanded form:

$$QaPbQcPdQeQ = 1, \qquad QaPc^{-1}Qe^{-1}Q = 1.$$

From these equations we see that the surface has two polygons (one for each equation), five edges (a, b, c, d, e), and two vertices (P and Q). Figure 2.25 shows how the polygons fit together at Q.

From the procedure for identifying vertices we see that a vertex is an endpoint of two or more singly appearing edges, either twice or not at all. It is possible, of course, that the vertex appears twice as an endpoint of the same edge—once as initial point and once as terminal point. We call the edges that occur only once boundary edges. Because the boundary edges are always in pairs at each vertex, they may be arranged into cyclic sets to form boundary curves. We shall show that the number of boundary curves is invariant under combinatorial equivalence and that this number equals the number of cuffs in the canonical equation for the surface.

In Operation 1 an equation of the form $ABC = 1$ is replaced by a pair of equations $AxC = 1$ and $x = B$. Because the new edge x appears twice, it is not a boundary edge. The second equation shows that the edge x and the block B have the same initial and terminal vertices. By applying this fact in the first equation we see that the terminal vertex of block A is the initial vertex of block B and the terminal vertex of B is the initial vertex of C. Thus adding the nonboundary edge x creates no vertices and does not change the vertex identification of the original system that contained the equation $ABC = 1$. This means that Operation 1 cannot change the number of boundary curves.

Operation 2 replaces an equation $AB = 1$ by $Axx^{-1}B = 1$. The new edge x is not a boundary edge. Because the vertex that starts x finishes x^{-1},

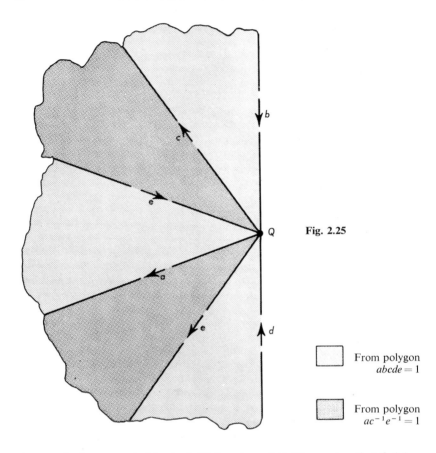

Fig. 2.25

From polygon
$abcde = 1$

From polygon
$ac^{-1}e^{-1} = 1$

the terminal vertex of A is the initial vertex of B. The vertex that finishes x and starts x^{-1} can occur only between x and x^{-1}. This new vertex, which meets no boundary edges, cannot affect the number of boundary curves. Because the identifications among the old vertices have not changed, we conclude that the number of boundary curves is invariant under Operation 2. Under Operation 3 a single edge x is replaced by a pair of consecutive edges uv. If x is not a boundary edge, replacement will not change the boundary edges and their vertices, If x is a boundary edge, the pair of boundary edges u and v will replace x in one of the boundary curves. Again, the number of boundary curves is invariant. Because Operations $1'$, $2'$, and $3'$ are inverse to Operations 1, 2, and 3, the number of boundary curves is invariant under the three remaining types of combinatorial equivalence.

Let us now write our two types of canonical form by using both edge and vertex symbols.

$$Pc_1Pc_1P \cdots Pc_qPc_qPd_1Q_1e_1Q_1d_1^{-1}P \cdots Pd_rQ_re_rQ_rd_r^{-1}P = 1,$$

$$Pa_1Pb_1Pa_1^{-1}Pb_1^{-1}P \cdots Pa_pPb_pPa_p^{-1}Pb_p^{-1}Pd_1Q_1e_1Q_1d_1^{-1}P$$

$$\cdots Pd_rQ_re_rQ_rd_r^{-1}P = 1.$$

In both equations each of the boundary edges $e_1, \ldots, e_r$ is a boundary curve. Thus, the number of cuffs equals the number of boundary curves.

We shall show the combinatorial invariance of the genus p or q by first showing the invariance of the Euler characteristic. The Euler characteristic χ of a surface is defined by $\chi = n_0 - n_1 + n_2$, where n_0 is the number of vertices, n_1, the number of edges, and n_2, the number of polygons. Operation 1 adds one equation and one edge but no vertices. Thus n_1 and n_2 are both increased by one, but χ is unchanged. Operations 2 and 3 both increase n_0 and n_1 by one but do not change n_2. Again χ is unchanged. This shows that the Euler characteristic is a combinatorial invariant. The expanded forms of the canonical equations show that

$$\chi = (1 + r) - (q + 2r) + 1 = 2 - q - r \qquad \text{Non orientable}$$

for a nonorientable surface and

$$\chi = (1 + r) - (2p + 2r) + 1 = 2 - 2p - r \qquad \text{orientable.}$$

for an orientable surface. Because χ, r, and orientability are all combinatorial invariants, the genus is also a combinatorial invariant.

We have accomplished our goal of proving the combinatorial invariance of our canonical equation. We now know that no one surface can be reduced by two different sequences of combinatorial equivalences to give canonical equations with different values of p, q, and r. As a by-product we have discovered simple procedures to determine orientability, genus, and number of cuffs. From the invariants we can write down the canonical equation of a surface without the tedious procedure used to derive the canonical equation. The original derivation was necessary to prove the existence of a canonical equation, but the present methods are more efficient for finding the equation.

2.5   Topological Surfaces

We have just seen that a surface is completely classified under the relation of combinatorial equivalence by its orientability or nonorientability, genus, and number of cuffs. There is, therefore, a wide variety of possible surfaces, which is all the more remarkable when we consider that from a "local" or "myopic" point of view all surfaces are more or less the same.

Let us make this last statement precise. A *neighborhood* of a point x on a surface S is any set of points of S that contains, for some number (possibly very small) $\delta > 0$, all points of S whose distance from x is $< \delta$. Then, no matter what the surface S and the point x in S, there is some neighborhood of x in S that is homeomorphic to a disk (why?). If x is an interior point of S, there will be a homeomorphism of an appropriate neighborhood of x in S with a disk that takes x onto the center of the disk. If x is a boundary point of S, there will be a homeomorphism of some neighborhood of x in S with a disk that takes x onto a boundary point of the disk. In this sense there are only two species of points on S—interior points and boundary points. It is visually obvious, though somewhat sticky to prove, that interior points and boundary points are distinct from one another. We take this for granted but shall not give the proof.

A *topological surface S* is a point set in some Euclidean space with the property that each point x of S has some neighborhood that is homeomorphic to a disk. As above, we may distinguish between the interior and boundary points of S. Every surface, in the original sense of Section 2.1, is also a topological surface. It is natural to ask, "Is every topological surface also a surface in our original sense?" That is, can every topological surface be subdivided into a finite number of polygons with the usual intersection requirements of Section 2.1? A little thought convinces us that the answer is an immediate NO unless we restrict the question somewhat; for example, the ordinary plane (two-dimensional Euclidean space) satisfies the definition of a topological surface but cannot be subdivided into a *finite* number of polygons. It can be subdivided into an *infinite* number of polygons, but this is not permitted by our original definition.

A good way to restrict the question is to require that the topological surface S be closed and bounded in whatever Euclidean space it lies. The problem of deciding whether S is a surface in our original sense is known as the *triangulation problem for surfaces*. The word "triangulation" is used because any surface that can be subdivided into polygons can also be further subdivided into triangles. The triangulation problem for surfaces was solved in the *affirmative* by T. Rado in 1925. The triangulation problem for three-dimensional manifolds (the three-dimensional analogue of a surface) was solved by E. E. Moise in 1952 and by R. H. Bing (in a more general sense) in 1954. In dimensions higher than three the triangulation problem is still an outstanding unsolved problem of topology.

Closely related to the triangulation problem is another problem best known by its German name: the *Hauptvermutung* (principal conjecture). Stated for surfaces, it is the following. Suppose we have two closed and bounded surfaces, S and S', in Euclidean spaces. Then we know from the solution to the triangulation problem that each can be subdivided into polygons in an admissible way. Suppose that S and S' are homeomorphic as point sets. Does it follow that the subdivisions of S and S' are combinatorially equivalent in the sense of this chapter? The answer is YES. The actual argument is not hard to give on the basis of the material we have already developed, but it does need a little background in point set topology. The idea of the argument is simply to prove that no two surfaces with different canonical equations could be homeomorphic. The reader who knows a little point set topology may like to attempt the proof.

Like the triangulation problem, the Hauptvermutung is also solved affirmatively in dimension three but is unsolved beyond that. A very general version of the Hauptvermutung for complexes was shown to be false by John Milnor in 1961.

EXERCISES

Section 2.1

1. Which of the following systems of equations represent a surface? If the equations do not represent a surface, what condition is violated? If the equations do represent a surface, is the surface closed?

    (a) $abcd = 1$,
         $bd = ef$,
         $a^{-1}c = eb$.

    (b) $abcd = 1$,
         $bd = ac$,
         $efgh = 1$,
         $eg = fh$.

    (c) $abcd = 1$,
         $bd = ef$,
         $ac = fe$.

    (d) $abcd = 1$,
         $b = ac$.

2. For each of the following sets of conditions consider the corresponding locus in three-dimensional Euclidean space. In each case explain why the locus is not a surface.

    (a) $xyz = 0$, $x^2 + y^2 + z^2 \leq 1$.
    (b) $x^2 + y^2 = z^2$, $x^2 + y^2 + z^2 \leq 1$.
    (c) $[(x - 2)^2 + y^2 + z^2 - 1][(x + 2)^2 + y^2 + z^2 - 1] = 0$.

3. By trivial equivalences transform the system of equations on the left into the system of equations on the right.

(a) $ab^{-1}cd^{-1} = 1,$ $\qquad\qquad\qquad c = ba^{-1}d,$

$\quad a^{-1}ecb = 1.$ $\qquad\qquad\qquad\quad e = ab^{-1}c^{-1}.$

(b) $aba^{-1}cd^{-1}ebcf = 1,$ $\qquad\qquad c^{-1}dce^{-1}a^{-1}f^{-1}de^{-1}b = 1.$

(c) $abcd = 1,$ $\qquad\qquad\qquad\qquad xy = zw,$

$\quad aed^{-1}c = 1.$ $\qquad\qquad\qquad\quad zw^{-1} = tx^{-1}.$

4. List all equations into which $abb = 1$ can be transformed by a sequence of transpositions.

5. A function $f(x, y)$ with points in three-dimensional Euclidean space as values is defined over the disk $x^2 + y^2 \leq 1$ by the formula

$$f(x, y) = (|x|, y, (x^2 + y^2 - 1)x).$$

Give an edge equation for the quadrilateral determined by this function. What surface is formed by the points of the quadrilateral?

6. A function $f(x, y)$ with points in five-dimensional Euclidean space as values is defined over the disk $x^2 + y^2 \leq 1$ by the formula

$$f(x, y) = [|x|, |y|, (xy - |xy|)y, (x^2 + y^2 - 1)x, (x^2 + y^2 - 1)y].$$

Give an edge equation for the quadrilateral determined by this function. What surface is formed by the points of the quadrilateral?

Section 2.2

In Chapter 1 "cut-and-paste" operations were used informally. In the exercises that follow justify these informal operations by showing algebraically that the system of equations on the left is combinatorially equivalent to that on the right. The references are to figures in Chapter 1.

1. $abb^{-1}a^{-1} = 1$(Figure 1.9), $\qquad\qquad aa^{-1} = 1$(Figure 1.18).

2. $abca^{-1}b^{-1}c^{-1} = 1$(Figure 1.21), $\qquad\quad efe^{-1}f^{-1} = 1$(Figure 1.22).

3. $a_1b_1^{-1}a_2^{-1}b_1 = 1,$
   $a_2b_2^{-1}a_3^{-1}b_2 = 1$(Figure 1.26), $\qquad\quad aba^{-1}b^{-1} = 1.$
   $a_3b_3^{-1}a_1^{-1}b_3 = 1,$

4. $a_3b_1a_1^{-1}b_3^{-1} = 1,$
   $a_1b_2a_3^{-1}b_4^{-1} = 1,$
   $a_4b_3a_2^{-1}b_1^{-1} = 1,$ (Figure 1.27) $\qquad aba^{-1}b^{-1} = 1.$
   $a_2b_4a_4^{-1}b_2^{-1} = 1,$

5. $abac^{-1} = 1$(Figure 1.31), $ddb^{-1}c^{-1} = 1$(Figure (1.32).

6.   I $a_1 d_1 e_1^{-1} b_2 = 1$, $b_1 b_2 a_1 d_1 d_2 a_2 = 1$,

    II $e_1 d_2 a_2 b_1 = 1$,

    III $a_2^{-1} d_1 e_2 c_2 = 1$, (Figure 1.39 or 1.38) $c_1 c_2 a_2^{-1} d_1 d_2 a_1^{-1} = 1$

    IV $e_2^{-1} d_2 a_1^{-1} c_1 = 1$,   (Figure 1.40).

7. $abab = 1$ (Figure 1.46), $cc = 1$(Figure 1.51 or 1.50).

Section 2.3

1. Reduce each system of equations to a single canonical equation.

(a) $abcdef = 1$, $a^{-1}cegbd = 1$.

(b) $abcd = 1$, $eafc = 1$, $gfhe = 1$, $hdgb = 1$.

(c) $acba^{-1} db^{-1} = 1$, $ecef = 1$.

2. Show that every surface can be triangulated.

CAUTION. The subdivisions of a torus into triangles shown in Figures 2.26 and 2.27 are not triangulations. Why?

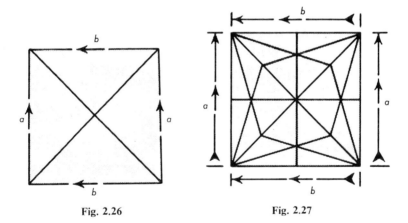

Fig. 2.26   Fig. 2.27

HINT. Figure 2.27 is obtained from Figure 2.26 by drawing the medians of all the triangles of Figure 2.26. If the same process is applied to Figure 2.27 instead of 2.26, the resulting subdivision is a triangulation.

Section 2.4

1. Without deriving the canonical equations, classify the surfaces represented as follows:

(a) $abcde = 1$, $ad^{-1}be^{-1}c = 1$.
(b) $abcde = 1$, $bdfh = 1$, $aef^{-1} = 1$.
(c) $a_1 a_2 \cdots a_n a_1^{-1} a_2^{-1} \cdots a_n^{-1} = 1$.

(distinguish between n even and n odd)

2. Δ_1: 124, 235, 346, 457, 561, 672, 713
       134, 245, 356, 467, 571, 612, 723
 Δ_2: 123, 156, 167, 172, 256, 268, 275, 283
       341, 357, 374, 385, 451, 468, 476, 485
What surfaces do Δ_1 and Δ_2 triangulate?

3. List all surfaces with non-negative Euler characteristics.

4. Let x, y, z be Cartesian coordinates in three-dimensional Euclidean space and let S be the locus of the equation

$$z^2 = [(x-2)^2 + y^2 - 1][(x+2)^2 + y^2 - 1](16 - x^2 - y^2).$$

What is the Euler characteristic of S? Is S orientable? Describe S. Write a canonical edge-equation for S.

5. A surface S is cut along k nonintersecting simple closed curves C_1, C_2, ..., C_k to form n surfaces $S_1, S_2, \ldots, S_n$.

(a) Can $n > k + 1$? Can $n < k + 1$? Why?
(b) Can any or all of $S_1, \ldots, S_n$ be nonorientable if S is orientable? Can any or all of $S_1, \ldots, S_n$ be orientable if S is nonorientable? If either answer is yes, give an example.
(c) Show that the Euler characteristic of S is the sum of the Euler characteristics of $S_1, \ldots, S_n$.
HINT: If a simple closed curve is divided by vertices into edges, the number of vertices equals the number of edges. Compare the polygons, vertices, and edges of $S_1, \ldots, S_n$ as separate surfaces with the polygons, vertices, and edges of $S_1, \ldots, S_n$ as parts of S.

3 COMPLEX CONICS AND COVERING SURFACES

3.1 Complex Conics

In a beginning study of analytical geometry loci of equations of the form

$$Q(x, y) = 0,$$

where

$$Q(x, y) = Ax^2 + Bxy + Cy^2 + Dx + Ey + F,$$

are classified as hyperbolas, parabolas, ellipses (including circles as special cases), and degenerate loci. These curves were named *conic sections* or simply *conics* because ellipses, parabolas, hyperbolas, and most of the degenerate loci may be obtained as the intersection of a cone and a plane. A hyperbola with two unbounded branches, an ellipse with a single bounded branch, a parabola with a single unbounded branch, and the empty locus seem quite unrelated. We find a partial explanation of these differences when we attempt to solve the equation

$$Cy^2 + (Bx + E)y + (Ax^2 + Dx + F) = 0$$

for y in terms of x. For a given value of x there are points (x, y) on the locus only if this quadratic equation for y has real roots. If the locus is an ellipse, there is a finite interval of values of x for which there are real roots; if the locus is empty, there are no such values of x. For a hyperbola the set of values of x with points (x, y) on the conic is either the whole x-axis or the x-axis with a finite interval deleted (Figures 3.1 and 3.2).

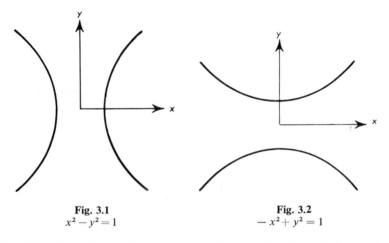

Fig. 3.1	Fig. 3.2
$x^2 - y^2 = 1$	$-x^2 + y^2 = 1$

The allowable values of x for a parabola are either the whole x-axis or half the x-axis; that is, a segment bounded in one direction and unbounded in the other (Figures 3.3 and 3.4). To eliminate some of the distinctions between the various conics we extend our perspective by replacing the real variables x and y with variables z and w which range over the extended complex number system, that is, over the Riemann sphere on which the special complex number ∞ is the north pole. The locus of the equation $Q(z, w) = 0$ is *degenerate* if the following conditions hold:

1. $Q(z, w)$ is not quadratic; that is, $A = B = C = 0$.
2. $Q(z, w)$ does not depend on both variables, that is $A = B = D = 0$ or $B = C = E = 0$.
3. $Q(z, w)$ factors into two linear polynomials.

We shall limit our study to nondegenerate complex conics. As a preliminary we review some properties of complex numbers. If $z = x + iy$,

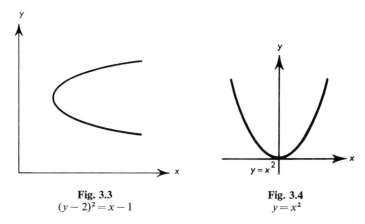

Fig. 3.3
$(y-2)^2 = x-1$

Fig. 3.4
$y = x^2$

the real numbers x and y are the *real* and *imaginary parts* of z and the x- and y-axes are the *real* and *imaginary axes* of the complex plane. If $z_1 = x_1 + iy_1$ and $z_2 = x_2 + iy_2$, the *sum* and *product* of z_1 and z_2 are $z_1 + z_2 = (x_1 + x_2) + i(y_1 + y_2)$,

$$z_1 z_2 = (x_1 + iy_1)(x_2 + iy_2) = x_1 x_2 + ix_1 y_2 + iy_1 x_2 + i^2 y_1 y_2$$

$$= (x_1 x_2 - y_1 y_2) + i(x_1 y_2 + y_1 x_2).$$

By rewriting z_1 and z_2 in polar form, we find

$$z_1 z_2 = r_1(\cos \theta_1 + i \sin \theta_1) r_2(\cos \theta_2 + i \sin \theta_2)$$

$$= r_1 r_2((\cos \theta_1 \cos \theta_2 - \sin \theta_1 \sin \theta_2) + i(\cos \theta_1 \sin \theta_2 + \sin \theta_1 \cos \theta_2))$$

$$= r_1 r_2(\cos(\theta_1 + \theta_2) + i \sin (\theta_1 + \theta_2)).$$

Thus we can multiply complex numbers written in polar form by multiplying absolute values and adding arguments. In particular, if

$$z = r(\cos \theta + i \sin \theta),$$

$$z^2 = r^2(\cos 2\theta + i \sin 2\theta), \qquad z^n = r^n(\cos n\theta + i \sin n\theta),$$

$$\frac{1}{z} = \frac{1}{r}(\cos (-\theta) + i \sin (-\theta)) = \frac{1}{r}(\cos \theta - i \sin \theta).$$

To define $\sqrt{z}$ we examine the equation

$$w^2 = z, \qquad \text{where } w = s(\cos \phi + i \sin \phi).$$

When

$$s^2(\cos 2\phi + i \sin 2\phi) = r(\cos \theta + i \sin \theta),$$

$r = s^2$. Because the argument of a complex number is determined only up to addition of an integral multiple of 2π, we can conclude that

$$2\phi = \theta + k2\pi, \quad \text{for some integer } k,$$

or

$$\phi = \frac{\theta}{2} + k\pi.$$

Because values of ϕ which differ by multiples of 2π correspond to the same complex number w, $\phi = \theta/2$ and $\phi = \theta/2 + \pi$ determine the only distinct solutions of the equation $w^2 = z$. By restricting ϕ to the interval $-\pi/2 \leq \phi \leq 3\pi/2$ we have one solution w with $-\pi/2 \leq \phi \leq \pi/2$ and the second with $\pi/2 \leq \phi \leq 3\pi/2$. We use $\sqrt{z}$ to denote the first value of w. Because $\cos(\phi + \pi) = -\cos\phi$ and $\sin(\phi + \pi) = -\sin\phi$, the second value of w is $-\sqrt{z}$. These symbols are unambiguous unless $\phi = -\pi/2$, $\pi/2$, or $3\pi/2$. These values of ϕ correspond to the complex numbers z with $\theta = \pi$, which are the negative real numbers. Except for the negative half of the real axis, $\sqrt{z}$ is a continuous function of z throughout the complex plane. Although different definitions of $\sqrt{z}$ and $-\sqrt{z}$ would shift the location of the ambiguity, the ambiguity cannot be avoided. As z moves from the second quadrant to the third quadrant across the negative real axis, $\sqrt{z}$ must be replaced by $-\sqrt{z}$ if the solution of $w^2 = z$ is to be a continuous function of z. For $z = 0$, $\sqrt{z} = -\sqrt{z} = 0$. Also, for $z = \infty$, $\sqrt{z} = -\sqrt{z} = \infty$. The interpretation of $\sqrt{\infty}$ will be clear when we have learned how to test whether a point (z, ∞) with z finite, a point (∞, w) with w finite, or the point (∞, ∞) satisfies a quadratic equation $Q(z, w) = 0$.

We illustrate the procedure with the polynomials

$$Q_1(z, w) = w^2 - z, \qquad\qquad Q_2(z, w) = wz + z - w,$$

$$Q_3(z, w) = w^2 + wz - z, \qquad\qquad Q_4(z, w) = z^2 + w^2 + z.$$

The first step to test (∞, w) is to replace z by $1/z'$. The equations $Q_j(z, w) = 0$ become

$$w^2 - \frac{1}{z'} = 0, \qquad\qquad \frac{w}{z'} + \frac{1}{z'} - w = 0,$$

$$w^2 + \frac{w}{z'} - \frac{1}{z'} = 0, \qquad\qquad \frac{1}{z'^2} + w^2 + \frac{1}{z'} = 0.$$

Any solution $(z'\ w)$ determines a point $(z, w) = (1/z', w)$ on the original locus. For any finite nonzero value of z, hence for any nonzero finite value of z', the solutions for w do not change if the equation is multiplied by a power of z'. By multiplying by the lowest power of z' that will clear the denominators we derive the new equations

$$z'w^2 - 1 = 0, \qquad w + 1 - wz' = 0,$$

$$z'w^2 + w - 1 = 0, \qquad 1 + w^2z'^2 + z' = 0.$$

The point $(z, w) = (\infty, w)$ is *defined* to be on the original locus if $(z', w) = (0, w)$ satisfies the derived equation. Substitution of $z' = 0$ into these equations yields

$$-1 = 0, \qquad w + 1 = 0, \qquad w - 1 = 0, \qquad 1 = 0.$$

We have found that there are no points (∞, w) with w finite on the loci $Q_1(z, w) = 0$ or $Q_4(z, w) = 0$ but have shown that $(\infty, -1)$ is on the locus of $wz + z - w = 0$ and $(\infty, 1)$ is on the locus of $w^2 + wz - z = 0$.

To find the points (z, ∞) with z finite on the loci we replace w with $1/w'$, and multiply the equations by the power of w' that will just eliminate the denominators. The equations $Q_j(z, w) = 0$ become

$$1 - w'^2z = 0, \qquad z + w'z - 1 = 0,$$

$$1 + w'z - w'^2z = 0, \qquad w'^2z^2 + 1 + w'^2z = 0.$$

The solutions $(z, w') = (z, 0)$ determine the points (z, ∞) on the loci of $Q_j(z, w) = 0$. Substitution of $w' = 0$ in the four equations yields

$$1 = 0, \qquad z - 1 = 0, \qquad 1 = 0, \qquad 1 = 0,$$

so that the loci $Q_j(z, w) = 0$ for $j \neq 2$ have no points (z, ∞) with z finite. The conic $wz + z - w = 0$ contains the point $(1, \infty)$. To test whether (∞, ∞) is on the four loci, replace z by $1/z'$ and w by $1/w'$ and then clear fractions. The resulting equations are

$$z' - w'^2 = 0, \qquad 1 + w' - z' = 0,$$

$$z' + w' - w'^2 = 0, \qquad w'^2 + z'^2 + w'^2z' = 0.$$

The point (∞, ∞) is on the original locus if and only if $z' = w' = 0$ is a solution of the new equation. Thus (∞, ∞) is on the loci $Q_j(z, w) = 0$ for $j \neq 2$ but is not on the locus $Q_2(z, w) = 0$. Because (∞, ∞) is the only point (∞, w) that satisfies $w^2 - z = 0$, ∞ is the only square root of ∞.

Now that we have defined $\sqrt{z}$ throughout the extended complex plane we can use the traditional formula to express in terms of z the solutions w of the quadratic equation

$$Cw^2 + (Bz + E)w + (Az^2 + Dz + F) = 0.$$

The two values of the second coordinate of points (z, w) of the conic $Q(z, w) = 0$ are

$$w = \frac{-(Bz + E) \pm \sqrt{(Bz + E)^2 - 4C(Az^2 + Dz + F)}}{2C},$$

provided that $C \neq 0$.

As a preliminary to the general case, we consider the special situation in which $C = 0$. Under this assumption the conic consists of all points (z, w) with

$$w = -\frac{Az^2 + Dz + F}{Bz + E}.$$

Because the conic is nondegenerate, B and E cannot both be zero. A unique value of w is determined for every value of z except possibly $z = -E/B$ (if $B \neq 0$) and $z = \infty$. When $B \neq 0$, the linear polynomial $Bz + E$ cannot be a factor of $Az^2 + Dz + F$, for $Q(z, w)$ has no linear factors. A check shows that $(-E/B, \infty)$ is on the locus when $B \neq 0$. Additional calculation verifies the following:

1. (∞, ∞) is the only point (∞, w) on the locus if $A \neq 0$ or $A = B = 0$.
2. $(\infty, -D/B)$ is the only point (∞, w) if $A = 0$ but $B \neq 0$.

Because there is a unique point (z, w) on the locus for each value of z, z can be used as a single coordinate on the conic and the conic is topologically a sphere.

Returning to the general case, we rewrite the expression under the radical sign as

$$d(z) = (B^2 - 4AC)z^2 + (2BE - 4CD)z + (E^2 - 4CF).$$

If $B^2 - 4AC = 0$, this polynomial is linear instead of quadratic. In real analytical geometry we know that $B^2 - 4AC = 0$ means that the conic is a parabola. We postpone consideration of the complex parabola. If $d(z)$ were a perfect square $(Kz + H)^2$, the two values of w would be

$$w = \frac{-(Bz + E) \pm (Kz + H)}{2C}.$$

This would imply that

$$Q(z, w) = C\left(w + \frac{Bz + E}{2C} - \frac{Kz + H}{2C}\right)\left(w + \frac{Bz + E}{2C} + \frac{Kz + H}{2C}\right),$$

thus contradicting the nondegeneracy of the conic. Because $d(z)$ is not a perfect square, the equation $d(z) = 0$ has two distinct roots

$$t_1 = r_1 + is_1 \quad \text{and} \quad t_2 = r_2 + is_2$$

and

$$d(z) = (B^2 - 4AC)(z - t_1)(z - t_2).$$

The identity

$$\left(\sqrt{B^2 - 4AC}\sqrt{z - t_1}\sqrt{z - t_2}\right)^2 = (B^2 - 4AC)(z - t_1)(z - t_2)$$

implies that $\sqrt{d(z)}$ is either

$$\sqrt{B^2 - 4AC}\sqrt{z - t_1}\sqrt{z - t_2} \quad \text{or} \quad -\sqrt{B^2 - 4AC}\sqrt{z - t_1}\sqrt{z - t_2}.$$

If we set $G = \sqrt{B^2 - 4AC}$, the two solutions for w are

$$w_1 = \frac{-(Bz + E) + G\sqrt{z - t_1}\sqrt{z - t_2}}{2C},$$

$$w_2 = \frac{-(Bz + E) - G\sqrt{z - t_1}\sqrt{z - t_2}}{2C}.$$

Because $z - t_1$ is a negative real number only on the halfline defined by $y = s_1$ and $x < r_1$ and $z - t_2$ is a negative real number only on the halfline defined by $y = s_2$ and $x < r_2$, the product $\sqrt{z - t_1}\sqrt{z - t_2}$ is unambiguously defined throughout the complex plane except on these halflines.[†] Hence w_1 and w_2 are continuous functions of z whose domain (Figure 3.5) is the complex plane with two halflines deleted. Figure 3.6 shows the halflines as an edge a from ∞ to t_1 and an edge b from ∞ to t_2 on the Riemann sphere. With these edges the sphere is a quadrilateral with edge equation $aa^{-1}bb^{-1} = 1$. For all points except on a or b, z can be used as a single coordinate for the point (z, w_1) on the conic. The set of these points lies on one sheet of the conic. Similarly the points (z, w_2) lie on a second sheet. Considering the points (z, w_1) and (z, w_2) as the graph of w_1 and w_2 as functions of z, we say that the points (z, w_1) and (z, w_2) cover the point z. The graph is a two-sheeted covering of the Riemann sphere with the edges a and b deleted. Now $w_1 = w_2$ only at t_1, t_2, and ∞, the vertices of the excluded edges. Hence the sheets do not intersect.

---

† If $s_1 = s_2$, one halfline would be part of the other. The discussion of this special case differs from that of the general case but the principles and the results are the same.

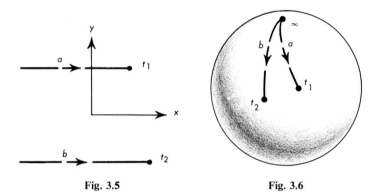

Fig. 3.5   Fig. 3.6

We bring the edges a and b back into the problem. As z crosses the edge a, the solutions w_1 and w_2 should be interchanged if they are to be continuous functions of z. In Figure 3.5, as z approaches a point on a from above, the point (z, w_1) approaches a different point on the conic from that reached when z approaches the edge a from below. When the first sheet is extended to cover a, two different symbols a_1 and a_2 should be used to represent the two different edges of the sheet that cover a. Similarly, two edges b_1 and b_2 cover b. An edge equation

$$a_1 a_2^{-1} b_1 b_2^{-1} = 1$$

for the first sheet may be derived from the equation

$$a a^{-1} b b^{-1} = 1$$

of the sphere by replacing a and b by the edges covering them. Because the point (z, w_1) changes to (z, w_2) as z crosses a or b, the corresponding equation for the second sheet is

$$a_2 a_1^{-1} b_2 b_1^{-1} = 1.$$

The surface defined by the equations

$$a_1 a_2^{-1} b_1 b_2^{-1} = 1 \quad \text{and} \quad a_2 a_1^{-1} b_2 b_1^{-1} = 1$$

is called the *Riemann surface* of $Q(z, w) = 0$, where the equation is considered as an implicit definition of w as a double-valued function of z.

If vertex symbols are inserted, the equations of the Riemann sphere and Riemann surface become

$$P a R a^{-1} P b S b^{-1} P = 1$$

and

$$P_1 a_1 R_1 a_2^{-1} P_2 b_1 S_1 b_2^{-1} P_1 = 1, \qquad P_2 a_2 R_1 a_1^{-1} P_1 b_2 S_1 b_1^{-1} P_2 = 1.$$

The points P, R, and $S(\infty, t_1,$ and t_2 on the Riemann sphere) are covered by two, one, and one points, respectively on the Riemann surface. Because $w_1 = w_2$ when $z = t_1$ or $z = t_2$, each of the points t_1 and t_2 is covered by exactly one point on the conic. Thus the conic and the Riemann surface both have a single point covering t_1 and a single point covering t_2. At R_1 and S_1 the two sheets of the Riemann surface come together so that R and S each have only one instead of two covering points. The points R_1 and S_1 are called *branch points* of the Riemann surface.

Because there are two polygons, four edges (a_1, a_2, b_1, b_2) and four vertices (P_1, P_2, R_1, S_1), the Riemann surface has Euler characteristic 2. This shows that the Riemann surface of $Q(z, w) = 0$ is a sphere.

To find the points on the conic covering $z = \infty$ we return to the equation

$$Cw^2 + (Bz + E)w + (Az^2 + Dz + F) = 0.$$

Assume $A \neq 0$. Replacing z by $1/z'$, and multiplying by z'^2, we derive

$$Cw^2z'^2 + (Bz' + Ez'^2)w + (A + Dz' + Fz'^2) = 0.$$

Setting $z' = 0$, we have $A = 0$. Hence there is no point (∞, w) with finite w on the conic. The two substitutions $z = 1/z'$ and $w = 1/w'$ lead to

$$Cz'^2 + (Bz' + Ez'^2)w' + (Aw'^2 + Dz'w'^2 + Fz'^2w'^2) = 0.$$

Because $z' = w' = 0$ is a solution, (∞, ∞) is on the conic.

If $A = 0$, the substitution $z = 1/z'$ leads to

$$Cw^2z' + (B + Ez')w + (D + Fz') = 0.$$

When $z' = 0$, this equation is $Bw + D = 0$. Because $A = 0$ and $B^2 - 4AC \neq 0$, $B \neq 0$. Therefore $(\infty, -D/B)$ is on the conic. Replacing w by $1/w'$, we derive

$$Cz' + (B + Ez')w' + (Dw'^2 + Fz'w'^2) = 0.$$

Again (∞, ∞) is on the conic.

If $A \neq 0$, P on the Riemann sphere is covered by the single point (∞, ∞) on the conic. If $A = 0$, the two distinct points $(\infty, -D/B)$ and (∞, ∞) cover P. When P is covered by two points on the conic, the conic and Riemann surface are topologically equivalent. When the conic has only one point covering P, the conic is equivalent to the Riemann surface, with the pair of points P_1 and P_2 identified.

We have almost completed the proof of the following theorem:

Theorem. A nondegenerate complex conic defined by a quadratic equation

$$Az^2 + Bzw + Cw^2 + Dz + Ew + F = 0$$

is topologically (a) a sphere if $A = 0$, if $C = 0$, or if $B^2 - 4AC = 0$, (b) a pinched sphere in all other cases.

In the case that remains to be studied $B^2 - 4AC = 0$. The polynomial $d(z)$ is

$$d(z) = (2BE - 4CD)z + (E^2 - 4CF).$$

Because the conic is nondegenerate, $2B - 4CD \neq 0$ and

$$d(z) = (2BE - 4CD)(z - t_1),$$

where

$$t_1 = \frac{E^2 - 4CF}{2BE - 4CD}.$$

Hence

$$w_1 = \frac{-(Bz + E) + L\sqrt{z - t_1}}{2C},$$

$$w_2 = \frac{-(Bz + E) - L\sqrt{z - t_1}}{2C},$$

where $L = \sqrt{2BE - 4CD}$. Because $\sqrt{z - t_1}$ is uniquely defined, except on the halfline determined by $y = s_1$ and $x < r_1$, w_1 and w_2 are continuous functions of z over the Riemann sphere with an edge a from t_1 to ∞ deleted. The edge equation $aa^{-1} = 1$ for the Riemann sphere leads to the equation $a_1 a_2^{-1} = 1$ for the sheet of the conic with points (z, w_1) and to $a_2 a_1^{-1} = 1$ for the sheet with points (z, w_2). From these equations we find that the Riemann surface is a sphere in which ∞ and t_1 are covered by a branch point instead of two distinct points. A check shows that $z = \infty$ is covered by the single point (∞, ∞) on the conic. When $B^2 - 4AC = 0$, the conic and the Riemann surface are topologically equivalent.

3.2   Covering Surfaces

In Section 3.1 we studied two-sheeted Riemann surfaces of complex quadratic equations. We now wish to extend our investigation to n-sheeted Riemann surfaces which are specified without reference to a polynomial equation. More generally, we wish to define the concept of an *n-sheeted covering surface* of any given surface. The covering surface is a Riemann surface if the covered surface is a sphere.

Starting with any set of equations representing a surface S, let us make n copies of the surface. We give equations for these copies by writing down the original set n times and adding a subscript to each edge symbol to denote the copy to which the edge belongs. If a letter occurs twice in the original set of equations, permute the subscripts on the second occurrences of the letter in the n copies. These permutations have the effect of interweaving the n copies to make an n-sheeted covering of the original surface. The only restriction on the permutations used is that the combined set of equations from all copies must be a combinatorial representation of some surface $\bar{S}$. Since the first condition for a combinatorial representation is automatically satisfied, only the second need be checked. The surface $\bar{S}$ is an n-sheeted covering surface of S.

To illustrate the interweaving process for forming covering surfaces we start with the equations

$$aab = 1, \qquad bcc = 1,$$

for a Klein bottle. We make three copies

$$a_1 a_1 b_1 = 1, \qquad b_1 c_1 c_1 = 1,$$
$$a_2 a_2 b_2 = 1, \qquad b_2 c_2 c_2 = 1,$$
$$a_3 a_3 b_3 = 1, \qquad b_3 c_3 c_3 = 1.$$

We interweave these copies by permuting the subscripts on the second occurrences of one or more of the letters a, b, and c in the three copies,

$$a_1 a_2 b_1 = 1, \qquad b_2 c_1 c_1 = 1,$$
$$a_2 a_3 b_2 = 1, \qquad b_1 c_2 c_2 = 1,$$
$$a_3 a_1 b_3 = 1, \qquad b_3 c_3 c_3 = 1.$$

In Chapter 2 we determined the vertex identification by finding all edges terminating in the same vertex. Our procedure actually gave the order in which the edges would be crossed on a path around the vertex. If the vertex is on a boundary curve, the sequence of edges starts and stops with a boundary edge. Otherwise, the sequence will repeat. As we follow the same procedure to determine the vertices of a covering surface, the sequence of edge symbols is the same except that subscripts are added to the edge symbols of the covered surface. Let P be a vertex of S and let P_1 be a vertex covering P. If P is on a boundary curve of S, the sequence of edges terminating at P_1 must have the same length because the edges of $\bar{S}$ covering boundary edges of S are boundary edges of $\bar{S}$. On the other hand, when P is not a boundary vertex, the sequence of edges about P_1 need not repeat so soon as the sequence about P because the sequence of edge symbols about P may be repeated many times before a combination of an edge symbol and a subscript repeats. The number of edges terminating at P_1 is k times the number of edges at P for some integer k. At P_1, k sheets of the covering surface are fitted together so that instead of one point over P in each of k sheets there is only the point P_1. When $k > 1$, P_1 is called a *branch point* of $\bar{S}$ as a covering of S and the deficiency $k - 1$ in the number of covering points is called the *order* of the branch point P_1.

Certain properties follow immediately from the definition of a covering surface. If the surface covered is orientable, its equation can be written so that no edge symbol will occur twice in the same sense. When n copies of these equations are interwoven to give an n-sheeted covering surface, each of the new edge symbols appears once in the expanded system of equations for each occurrence of the covered edge in the original equations. Furthermore, the covering edge appears in the same sense as the covered edge. This means that a covering surface of an orientable surface has a system of equations in which no edge symbol occurs twice in the same sense. Hence *every covering surface of an orientable surface is orientable*. Because every covering edge appears the same number of times as the covered edge, *a covering surface of a closed surface is closed*.

We now relate the Euler characteristics of a covering surface and the covered surface. Let $\bar{S}$ be an n-sheeted covering of a surface S formed from N_2 polygons, N_1 edges, and N_0 vertices. In $\bar{S}$ there are nN_2 polygons and nN_1 edges covering the polygons and edges of S but there are $nN_0 - \delta$ vertices; δ is the sum of the orders of the branch points on $\bar{S}$ as a covering of S. The Euler characteristic of S and $\bar{S}$ satisfy the relation $\chi_{\bar{S}} = n\chi_s - \delta$.

To complete the classification of $\bar{S}$ we must count the boundary curves of $\bar{S}$ if S is not closed and determine whether $\bar{S}$ is orientable. The question of orientability is already answered if S is orientable. Because the boundary curves of $\bar{S}$ cover those of S, there are at least as many boundary curves on $\bar{S}$ as on S and there are no more than n times as many boundary curves on $\bar{S}$ as on S. A single boundary curve of $\bar{S}$ may cover a boundary curve of S more than once, for it may bound more than one sheet.

Example 1. On the Riemann sphere let a be the real axis from 1 to 0, let b be the negative real axis from 0 to ∞, and let P, Q, and R be the points 1, 0, and ∞, respectively. Consider the Riemann surface of the equation

$$w^4 + (2 - 4z)w^2 + 1 = 0.$$

The four solutions of this equation for w in terms of z are

$$w_1 = \sqrt{z} + \sqrt{z - 1}, \qquad w_3 = -\sqrt{z} + \sqrt{z - 1},$$
$$w_2 = \sqrt{z} - \sqrt{z - 1}, \qquad w_4 = -\sqrt{z} - \sqrt{z - 1}.$$

If w is to vary continuously with z, w_1 must be interchanged with w_4 and w_2 with w_3 as z crosses b. On the other hand, w_1 must be interchanged with w_2 and w_3 with w_4 as z crosses a. If the jth sheet of the Riemann surfaces corresponds to the points (z, w_j), the four-sheeted Riemann surface has the following equations:

$$a_1 b_1 b_4^{-1} a_2^{-1} = 1,$$

$$a_2 b_2 b_3^{-1} a_1^{-1} = 1,$$

$$a_3 b_3 b_2^{-1} a_4^{-1} = 1,$$

$$a_4 b_4 b_1^{-1} a_3^{-1} = 1.$$

Rewritten with vertex symbols, the equations become

$$PaQbRb^{-1}Qa^{-1}P = 1$$

for the sphere and

$$P_1 a_1 Q_1 b_1 R_1 b_4^{-1} Q_2 a_2^{-1} P_1 = 1,$$

$$P_1 a_2 Q_2 b_2 R_2 b_3^{-1} Q_1 a_1^{-1} P_1 = 1,$$

$$P_2 a_3 Q_1 b_3 R_2 b_2^{-1} Q_2 a_4^{-1} P_2 = 1,$$

$$P_2 a_4 Q_2 b_4 R_1 b_1^{-1} Q_1 a_3^{-1} P_2 = 1,$$

for the Riemann surface. Each of the vertices P, Q, and R is covered by a pair of branch points of order 1. For the Riemann surface $\chi = 4(2) - 6 = 2$. This shows that the covering surface as well as the covered surface is a sphere.

Example 2. In Example 1 cut out a patch around the vertex P and consider the corresponding covering surface. The sphere with a patch removed has an equation $dbb^{-1}d^{-1}e = 1$, illustrated in Figures 3.7 and 3.8. The covering surface has the equations

$$S_1 d_1 Q_1 b_1 R_1 b_4^{-1} Q_2 d_2^{-1} S_2 e_1 S_1 = 1,$$

$$S_2 d_2 Q_2 b_2 R_2 b_3^{-1} Q_1 d_1^{-1} S_1 e_2 S_2 = 1,$$

$$S_3 d_3 Q_1 b_3 R_2 b_2^{-1} Q_2 d_4^{-1} S_4 e_3 S_3 = 1,$$

$$S_4 d_4 Q_2 b_4 R_1 b_1^{-1} Q_1 d_3^{-1} S_3 e_4 S_4 = 1.$$

The edge pairs $e_1 e_2$ and $e_3 e_4$ are the two boundary curves of the four-sheeted covering of the sphere with the one boundary curve e. If we let

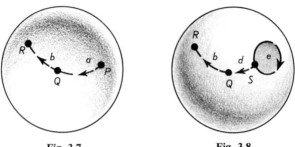

Fig. 3.7   Fig. 3.8

the curve e shrink to a point, we see that the branch points P_1 and P_2 in Example 2 are the limits of the curves $e_1 e_2$ and $e_3 e_4$, which doubly cover the curve e.

Of special interest are covering surfaces without branch points. If $n > 1$, there can be no unbranched n-sheeted coverings of the sphere, for the Euler characteristic of such a covering surface would have to be $2n$. The sum of the orders of the branch points of an n-sheeted covering of a sphere must be at least $2n - 2$.

We now show that every closed nonorientable surface has an unbranched two-sheeted covering which is orientable. In Chapter 2 we learned that every closed nonorientable surface may be represented by an equation

$$c_1 c_1 c_2 c_2 \cdots c_q c_q = 1.$$

By reversing Step 4 in the reduction to canonical form we may replace pairs of crosscaps with handles to transform this equation to one of two forms:

$$c_1 c_1 a_1 b_1 a_1^{-1} b_1^{-1} \cdots a_p b_p a_p^{-1} b_p^{-1} = 1$$

or

$$c_1 c_1 c_2 c_2 a_1 b_1 a_1^{-1} b_1^{-1} \cdots a_p b_p a_p^{-1} b_p^{-1} = 1.$$

In these forms we allow the case $p = 0$ in which there are no handles. The following pairs of equations give two-sheeted unbranched coverings in the two cases:

$$c_{11} c_{12} a_{11} b_{11} a_{11}^{-1} b_{11}^{-1} \cdots a_{p1} b_{p1} a_{p1}^{-1} b_{p1}^{-1} = 1,$$

$$c_{12} c_{11} a_{12} b_{12} a_{12}^{-1} b_{12}^{-1} \cdots a_{p2} b_{p2} a_{p2}^{-1} b_{p2}^{-1} = 1,$$

and

$$c_{11} c_{12} c_{21} c_{22} a_{11} b_{11} a_{11}^{-1} b_{11}^{-1} \cdots a_{p1} b_{p1} a_{p1}^{-1} b_{p1}^{-1} = 1,$$

$$c_{12} c_{11} c_{22} c_{21} a_{12} b_{12} a_{12}^{-1} b_{12}^{-1} \cdots a_{p2} b_{p2} a_{p2}^{-1} b_{p2}^{-1} = 1.$$

If the second equation in each of these pairs is inverted, the new pairs of equations will have every symbol appearing once in each sense. This shows that these covering surfaces are orientable. Because the covering is unbranched, the Euler characteristic of the orientable double covering is twice as large as that of the nonorientable surface. The projective plane, which has the equation $cc = 1$ and Euler characteristic 1, has the sphere as its orientable double covering.

In Chapter 1 we used a southern hemisphere with antipodal points on the equator identified as a model for the projective plane. If we complete the sphere with the understanding that each point in the northern hemisphere is identified with its antipodal point in the southern hemisphere, no new points but only new representations of old points have been added. This geometric model of the projective plane as a Euclidean sphere with antipodal points identified exhibits the sphere as a double covering of the projective plane. Lines through the center of the sphere intersect the sphere in pairs of antipodal points. If these lines rather than the pairs of antipodal points are thought of as "points" of a space, we find that the space of lines through a point in three-dimensional Euclidean space is topologically a projective plane. In Chapter 7 we shall study other spaces, called configuration spaces, in which lines, circles, or other geometric figures play the role of "points."

Example 3. On a torus $\bar{S}$, represented as a rectangle in the xy-plane with opposite edges identified, draw a regular subdivision into n hexagons so that each hexagon has two vertical edges, two edges with positive slope, and two edges with negative slope. Form a second torus S from a single hexagon by identifying opposite edges in the pattern $abca^{-1}b^{-1}c^{-1} = 1$. If each hexagon of $\bar{S}$ is considered as covering S with the vertical edges covering a, the edges of negative slope covering b, and the edges of positive slope covering c, $\bar{S}$ is an n-sheeted covering of S. Figures 3.9 and 3.10 illustrate this covering when $n = 4$. Because

$$0 = \chi_{\bar{s}} = n\chi_s - \delta = 0 - \delta,$$

$\bar{S}$ is an unbranched n-sheeted covering of the torus.

Example 4. Let x, y, z be Cartesian coordinates in a three-dimensional Euclidean space. Let surface S (Figure 3.11) be the locus of the equation

$$z^2 = (x^2 + y^2 - 1)[(x - 3)^2 + y^2 - 1][(x + 3)^2 + y^2 - 1](25 - x^2 - y^2).$$

Because z is real, the equation cannot be solved for z in terms of the point (x, y) of the xy-plane if $x^2 + y^2 < 1$, if $(x - 3)^2 + y^2 < 1$, if $(x + 3)^2 +$

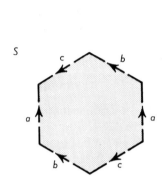

Fig. 3.9

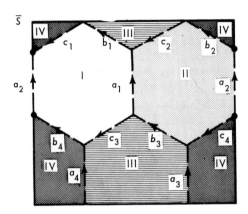

Fig. 3.10

$$\text{I} \quad a_1 b_1 c_1 a_2{}^{-1} b_4{}^{-1} c_3{}^{-1} = 1$$
$$\text{II} \quad a_2 b_2 c_2 a_1{}^{-1} b_3{}^{-1} c_4{}^{-1} = 1$$
$$\text{III} \quad a_3 b_3 c_3 a_4{}^{-1} b_1{}^{-1} c_2{}^{-1} = 1$$
$$\text{IV} \quad a_4 b_4 c_4 a_3{}^{-1} b_2{}^{-1} c_1{}^{-1} = 1$$

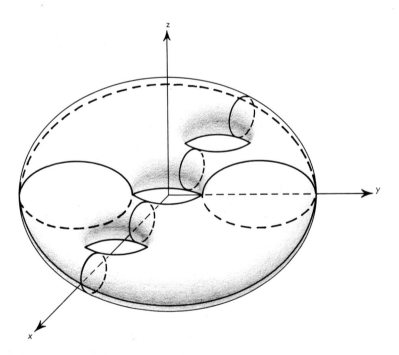

Fig. 3.11

$y^2 < 1$, or if $x^2 + y^2 > 25$. When (x, y) is outside the circles $x^2 + y^2 = 1$, $(x - 3)^2 + y^2 = 1$, and $(x + 3)^2 + y^2 = 1$, but inside the circle $x^2 + y^2 = 25$; there are two values of z, one positive and one negative, corresponding to (x, y). If (x, y) is on one of the four circles, $z = 0$. Thus S is a pretzel with three holes and is a closed orientable surface of genus 3. This description could be verified algebraically by dividing S into four polygons by drawing the curves in which S intersects the planes $y = 0$ and $z = 0$ and then studying the edge equations of these four polygons.

Let S_x, S_y, S_z, S_0 be the closed surfaces obtained from S by identifying points by symmetry in the x-axis [that is, the identification

$$(x, y, z) \leftrightarrow (x, -y, -z)],$$

in the y-axis, in the z-axis, and in the origin [that is,

$$(x, y, z) \leftrightarrow (-x, -y, -z)],$$

respectively; S is a two-sheeted covering of each of the surfaces S_x, S_y, S_z, S_0.

When S is considered as a covering of S_x, each of the eight points $(5, 0, 0)$, $(-5, 0, 0)$, $(4, 0, 0)$, $(-4, 0, 0)$, $(2, 0, 0)$, $(-2, 0, 0)$, $(1, 0, 0)$, $(-1, 0, 0)$ on the x-axis is the only point on S covering the corresponding point on S_x. Therefore S, as a two-sheeted covering of S_x, has eight branch points of order 1. Because $\chi_S = -4$ and $\chi_S = 2\chi_{S_x} - 8$, $\chi_{S_x} = 2$ and S_x is a sphere.

When S is considered as a covering of S_y, the branch points are $(0, 5, 0)$, $(0, -5, 0)$, $(0, 1, 0)$, $(0, -1, 0)$. Hence $2\chi_{S_y} - 4 = -4$ and $\chi_{S_y} = 0$. This means that S_y is either a torus or a Klein bottle. To determine whether S_y is orientable, we first subdivide S into four polygons by drawing the curves in which the planes $y = 0$ and $z = 0$ intersect S. Figure 3.12 shows the top two polygons as seen from above. All of the edges except e_1, e_2, e_3, and e_4 are in the plane $z = 0$, hence are on both the upper and lower halves of S. Let f_1, f_2, f_3, and f_4 be the edges on the lower half directly below e_1, e_2, e_3, and e_4. If III is under I and IV is under II, the following are edge equations for the four polygons:

I $\quad d_4 d_1 e_4^{-1} c_1 e_3^{-1} b_{11} b_{12} e_2^{-1} a_1 e_1^{-1} = 1$,

II $\quad d_3 d_2 e_1 a_2 e_2 b_{21} b_{22} e_3 c_2 e_4 = 1$,

III $\quad d_4 d_1 f_4^{-1} c_1 f_3^{-1} b_{11} b_{12} f_2^{-1} a_1 f_1^{-1} = 1$,

IV $\quad d_3 d_2 f_1 a_2 f_2 b_{21} b_{22} f_3 c_2 f_4 = 1$.

Because $(x, y, z) \leftrightarrow (-x, y, -z)$ identifies a point on the top of S with a point on the bottom, every point in S_y is represented by a single point on

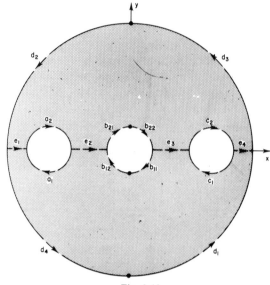

Fig. 3.12

the top of S except for double representation by points on the edges, where the top joins the bottom. These double representations may be eliminated by identifying a_2 with c_2^{-1}, b_{21} with b_{22}^{-1}, c_1 with a_1^{-1}, and d_2 with d_3^{-1}. Also d_4 with d_1^{-1} and b_{12} with b_{11}^{-1}. By making these substitutions in equations I and II we derive the following pair of equations for S_y:

$$d_1^{-1} d_1 e_4^{-1} a_1^{-1} e_3^{-1} b_{11} b_{11}^{-1} e_2^{-1} a_1 e_1^{-1} = 1$$

$$d_2^{-1} d_2 e_1 c_2^{-1} e_2 b_{22}^{-1} b_{22} e_3 c_2 e_4 = 1$$

Because every symbol occurs once in each sense, S_y is orientable. Hence S_y is a torus.

When S is considered as a covering for S_z, there are no branch points, for the z-axis does not intersect S. Hence $\chi_{S_z} = -2$ and S_z is either a sphere with two handles or a sphere with four crosscaps. The identification $(x, y, z) \leftrightarrow (-x, -y, z)$ matches each point on the back half of S with a point on the front half and identifies e_4 with e_1^{-1}, e_3 with e_2^{-1}, f_4 with f_1^{-1}, and f_3 with f_2^{-1}. Using these edge identities, we derive the following equations of S_z from I and III:

$$d_4 d_1 e_1 c_1 e_2 b_{11} b_{12} e_2^{-1} a_1 e_1^{-1} = 1$$

$$d_4 d_1 f_1 c_1 f_2 b_{11} b_{12} f_2^{-1} a_1 f_1^{-1} = 1$$

If the second equation is inverted, we see that S_z is orientable, hence a sphere with two handles.

Considering S as a covering for S_0, we again find no branch points, so that S_0 is either a sphere with two handles or a sphere with four crosscaps. The identification $(x, y, z) \leftrightarrow (-x, -y, -z)$ pairs a point on the top of S with one on the bottom and matches a_1 with c_2, b_1 with b_2, c_1 with a_2, d_3 with d_4, and d_1 with d_2. Starting from I and II, we find that S_0 has the edge equations

$$d_4 d_1 e_4^{-1} c_1 e_3^{-1} b_{11} b_{12} e_2^{-1} a_1 e_1^{-1} = 1$$

$$d_4 d_1 e_1 c_1 e_2 b_{11} b_{12} e_3 a_1 e_4 = 1$$

Because a_1 and e_1 have the opposite sense in one equation and the same sense in the other, either a_1 or e_1 will occur twice in the same sense no matter which equations are inverted. Thus S_0 is nonorientable and a sphere with four crosscaps.

3.3   Some Additional Examples of Riemann Surfaces

Example 5. Consider the Riemann surface of the equation

$$w^2 = (z - r_1)(z - r_2) \cdots (z - r_n), \tag{A}$$

where $r_1 > r_2 > \cdots > r_n$ are n distinct real numbers. The two solutions of this equation are

$$w_1 = \sqrt{z - r_1} \sqrt{z - r_2} \cdots \sqrt{z - r_n},$$

$$w_2 = -\sqrt{z - r_1} \sqrt{z - r_2} \cdots \sqrt{z - r_n}.$$

These definitions are unambiguous if z is not on the real axis to the left of r_1. If a_j is the segment of the real axis from r_j to r_{j+1} and a_n is the segment of the real axis to the left of r_n, the equation

$$a_1 a_2 \cdots a_n a_n^{-1} \cdots a_2^{-1} a_1^{-1} = 1$$

is an equation for the Riemann sphere. If z crosses the real axis at a point r, all factors $\sqrt{z - r_j}$ with $r_j > r$ change sign, whereas all other factors are unchanged. When z crosses a_j, w_1 and w_2 are interchanged if j is odd and not interchanged if j is even. Therefore the Riemann surface has the equations

$$a_{11} a_{21} a_{31} a_{41} \cdots a_{41}^{-1} a_{32}^{-1} a_{21}^{-1} a_{12}^{-1} = 1,$$

$$a_{12} a_{22} a_{32} a_{42} \cdots a_{42}^{-1} a_{31}^{-1} a_{22}^{-1} a_{11}^{-1} = 1.$$

When j is odd, a_{j1} and a_{j2}^{-1} are in the first equation, whereas a_{j2} and a_{j1}^{-1} are in the second. When j is even, a_{j1} and a_{j1}^{-1} are in the first, whereas a_{j2} and a_{j2}^{-1} are in the second.

When n is even, the middle pair of symbols is $a_{n1}a_{n1}^{-1}$ in the first equation and $a_{n2}a_{n2}^{-1}$ in the second. By combinatorial equivalence we can delete these pairs of symbols, thereby eliminating two edges and two vertices. The surface has two points, both covering ∞, but the locus had only one point covering ∞. This is another example in which the Riemann surface of an equation is topologically different from the locus of the equation. The locus is obtained from the Riemann surface by identifying the two points on the surface covering ∞. After the middle pairs have been deleted the equations for the Riemann surface are those that correspond to equation (A) when n is replaced by $n - 1$. Thus we have found that when the number of factors $z - r_i$ is even the Riemann surface is topologically unchanged if the last factor is omitted. We may therefore assume that n is odd and that the middle pair of symbols is $a_{n1}a_{n2}^{-1}$ in the first equation and $a_{n2}a_{n1}^{-1}$ in the second.

Since a Riemann surface is closed and orientable, the surface can be classified as soon as the genus or the Euler characteristic is calculated. We shall now find the branch points. Let P be the vertex between a_{11} and a_{21} in the first equation. It must be between a_{21}^{-1} and a_{12}^{-1} in the first equation and between a_{12} and a_{22} and a_{22}^{-1} and a_{11}^{-1} in the second. The single point P is the only point on the Riemann surface covering the point r_2, which is between a_1 and a_2 and between a_2^{-1} and a_1^{-1}. A similar calculation will show that each of the points $r_1, r_2, \ldots, r_n, r_{n+1} = \infty$ (n odd) is covered by a single point on the Riemann surface. Thus there are $n + 1$ branch points and the Euler characteristic is $\chi = 2(2) - (n + 1)$. Since $\chi = 2 - 2p$, the genus of the Riemann surface is $(n - 1)/2$ when n is odd. When n is even, the genus is $[(n - 1) - 1]/2 = n/2 - 1$. We have shown that the Riemann surface associated with the equation

$$w^2 = (z - r_1)(z - r_2) \cdots (z - r_n)$$

is a sphere with p handles where $p = n/2 - 1$ if n is even and $p = (n - 1)/2$ if n is odd. Because n can be any positive integer, p can be any non-negative integer. This means that every closed orientable surface is the Riemann surface of some double-valued function w defined by equation (A) for some value of n. This Riemann surface has branch points of order 1 covering $r_1, \ldots, r_n$ (n even) or $r_1, \ldots, r_n, r_{n+1} = \infty$ (n odd).

Example 6. For the examples up to this point we have used $\sqrt{z}$ but not higher roots. The equation $w^n = z$ has n roots which differ in their arguments by integral multiples of $2\pi/n$. These angles arise from division by n of values for arg z which differ by multiples of 2π. Among the nth roots of z, one denoted by $\sqrt[n]{z}$ can be selected with $-\pi/n \leq \arg\sqrt[n]{z} \leq \pi/n$. The nth roots of w then have the form

$$w_j = \left(\cos\frac{j2\pi}{n} + i\sin\frac{j2\pi}{n}\right)\sqrt[n]{z}, \qquad j = 0, 1, 2, \ldots, n-1.$$

For w to be a continuous function of z when z crosses the negative real axis, the subscript of w_j should be increased by one as z moves from the second to the third quadrant and decreased by one for motion in the reverse direction. If $j = n-1$, an increase by one should be interpreted as a return to zero. If a is the negative real axis from 0 to ∞, the Riemann sphere has equation $aa^{-1} = 1$ and the n-sheeted Riemann surface $\bar{S}$ of $w^n = z$ has the edge equations

$$a_0 a_{n-1}^{-1} = 1, \qquad a_1 a_0^{-1} = 1, \qquad a_2 a_1^{-1} = 1, \ldots, a_{n-1}a_{n-2}^{-1} = 1.$$

A check of the vertices on $\bar{S}$ shows that 0 and ∞ are each covered by a single vertex that is a branch point of order $n-1$. Hence

$$\chi_{\bar{S}} = n(2) - 2(n-1) = 2,$$

so that $\bar{S}$ is a sphere. In this example the locus of the equation is topologically equivalent to the Riemann surface.

Example 7. On the Riemann sphere let a be the real axis from 1 to 0 and let b be the negative real axis from 0 to ∞. The Riemann surface associated with the equation

$$w^3 = z(z-1)$$

has the equations

$$a_1 b_1 b_3^{-1} a_2^{-1} = 1,$$

$$a_2 b_2 b_1^{-1} a_3^{-1} = 1,$$

$$a_3 b_3 b_2^{-1} a_1^{-1} = 1.$$

To derive these equations define w for points on the first sheet by

$$w_1 = \sqrt[3]{z}\sqrt[3]{z-1},$$

where the arguments of these cube roots lie between $\pm\pi/3$.
For the second and third sheets define

$$w_2 = (\cos \pi/3 + i\sin \pi/3)w_1, \qquad w_3 = (\cos 2\pi/3 + i\sin 2\pi/3)w_1.$$

With these definitions the equations above follow immediately. Let P, Q, and R denote the points 1, 0, and ∞ on the sphere. We find that the vertex P_1, which is the initial point of a_1, is also the initial point of a_2 and a_3. Thus P_1 is a branch point of order 2 over the point P. The vertex R, which ends b, is covered by a branch point R_1, which ends b_1, b_2, b_3. Similarly, vertex Q between terminating a and starting b is covered by a single point R_1 which terminates a_1, a_2, a_3 and starts b_1, b_2, b_3. Because $\delta = 6$, the Euler characteristic of the Riemann surface is $\chi = 3(2) - 3(2) = 0$. Because a covering of a sphere must be closed and orientable, we conclude that the Riemann surface is a torus. In this example the Riemann surface and the locus are topologically equivalent.

EXERCISES

Section 3.1

1. Let x and y be Cartesian coordinates in the Euclidean plane. Classify each of the conics described by the following equations as an ellipse or circle, a hyperbola or pair of intersecting lines, a parabola, a point, or the empty locus:
   (a) $x^2 + xy + y^2 + 5 = 0$.
   (b) $x^2 + xy + y^2 - 5 = 0$.
   (c) $x^2 + xy + y^2 = 0$.
   (d) $x^2 + 2xy + y^2 + x = 0$.
   (e) $x^2 + 3xy + y^2 = 0$.

2. Let $f(z, w) = Az^2 + Bzw + Cw^2 + Dz + Ew + F$. If $C \neq 0$, show that $f(z, w)$ factors into two linear factors if and only if
$$CD^2 + B^2F + AE^2 - BDE - 4ACF = 0.$$

3. Find the points of the form (∞, w) on the complex conic defined by $w^2 - wz + z + 1 = 0$. For which values of z is there only one point on the conic of the form (z, w)?

4. Find the points of the form (∞, w) or (z, ∞) on the loci of these equations.
   (a) $(z^2 + 2z + 5)w^2 + (z + 2)w + 4z^2 + 3 = 0$.
   (b) $(z^2 + 3)w^2 + (z^3 + 2z)w + 2z^3 + z^2 + 3 = 0$.

5. What points z are covered by fewer than two points on the locus of $(z^2 + 1)w^2 + z^2w + \frac{1}{2}z^2 - \frac{1}{4} = 0$? For what values of z is the point (z, ∞) on the locus? What points cover $z = \infty$?

Section 3.2

1. What is the topological nature of the three-sheeted covering of a Klein bottle given in the text?

2. The surface $\bar{S}$ described by the equations

$$a_1b_1c_1d_1d_2^{-1}c_1^{-1}b_3^{-1}a_2^{-1} = 1,$$
$$a_2b_2c_2d_2d_3^{-1}c_3^{-1}b_2^{-1}a_1^{-1} = 1,$$
$$a_3b_3c_3d_3d_1^{-1}c_2^{-1}b_1^{-1}a_3^{-1} = 1$$

is a three-sheeted covering of the sphere S represented by

$$abcdd^{-1}c^{-1}b^{-1}a^{-1} = 1.$$

Which vertices of S are covered by fewer than three vertices of $\bar{S}$? What is the topological nature of $\bar{S}$?

3. The equations of a sphere S and a covering surface $\bar{S}$ are

$$S: aa^{-1}bb^{-1}cc^{-1}dd^{-1} = 1 \qquad \bar{S}: \begin{aligned} a_1a_2^{-1}b_1b_2^{-1}c_1c_2^{-1}d_1d_2^{-1} &= 1, \\ a_2a_1^{-1}b_2b_1^{-1}c_2c_1^{-1}d_2d_1^{-1} &= 1. \end{aligned}$$

How many branch points has $\bar{S}$ as a covering of S? What is the topological nature of $\bar{S}$?

4.
$$S: abcb^{-1}defe^{-1} = 1 \qquad T: \begin{aligned} a_1b_1c_1b_2^{-1}d_1e_1f_1e_1^{-1} &= 1, \\ a_2b_2c_2b_3^{-1}d_2e_2f_2e_3^{-1} &= 1, \\ a_3b_3c_3b_1^{-1}d_3e_3f_3e_2^{-1} &= 1. \end{aligned}$$

What is the topological nature of S? Has T any branch points as a three-sheeted covering of S? How many boundary curves (or cuffs) has T? What is the topological nature of T?

5.
$$S: aabbcc = 1 \qquad T: \begin{aligned} a_1a_2b_1b_3c_1c_1 &= 1, \\ a_2a_3b_2b_4c_2c_3 &= 1, \\ a_3a_4b_3b_1c_3c_4 &= 1, \\ a_4a_1b_4b_2c_4c_2 &= 1. \end{aligned}$$

Find the branch points and their orders for T as a four-sheeted covering of S. What is the topological nature of T?

6. What surfaces have closed orientable surfaces of genus 3 (spheres with three handles) as an unbranched covering?

7. Prove that a covering surface of a nonorientable surface is itself nonorientable if the number of sheets is odd.

8. For any positive integer n show that a Möbius band has an unbranched n-sheeted covering surface. When is this covering surface a Möbius band and when is it a cylinder?

9. Let x, y, z be Cartesian coordinates in Euclidean three-space. The surface S is the locus of the equation

$$z^2 = [(x - 2)^2 + y^2 - 1][(x + 2)^2 + y^2 - 1](16 - x^2 - y^2).$$

Let S_x and S_0 denote the closed surfaces obtained from S by identification of points by symmetry in the x-axis [that is,

$$(x, y, z) \leftrightarrow (x, -y, -z)]$$

and in the origin [that is,

$$(x, y, z) \leftrightarrow (-x, -y, -z)],$$

respectively. S is a two-sheeted covering of each of the surfaces S_x and S_0. Find the branch points of each covering. Determine the topological nature of each of the surfaces S, S_x, and S_0.

10.
$$S: a_1 b_1 c_1 b_1^{-1} a_2 b_2 c_2 b_2^{-1} a_3 b_3 c_3 b_3^{-1} = 1,$$
$$T: \begin{aligned} a_{11} b_{11} c_{11} b_{12}^{-1} a_{21} b_{21} c_{21} b_{2i_1}^{-1} a_{31} b_{31} c_{31} b_{3j_1}^{-1} &= 1, \\ a_{12} b_{12} c_{12} b_{13}^{-1} a_{22} b_{22} c_{22} b_{2i_2}^{-1} a_{32} b_{32} c_{32} b_{3j_2}^{-1} &= 1, \\ a_{13} b_{13} c_{13} b_{14}^{-1} a_{23} b_{23} c_{23} b_{2i_3}^{-1} a_{33} b_{33} c_{33} b_{3j_3}^{-1} &= 1, \\ a_{14} b_{14} c_{14} b_{11}^{-1} a_{24} b_{24} c_{24} b_{2i_4}^{-1} a_{34} b_{34} c_{34} b_{3j_4}^{-1} &= 1. \end{aligned}$$

Figure 3.13 represents S as a quadruply connected plane region. Show that T is an unbranched covering of S. Let the sphere S' and the surface T' be the closed surfaces obtained from S and T by "shrinking" each boundary curve of S or T to a point. Write equations for S' and T'. Let r be the number of boundary curves of T and let δ be the sum of the orders of the branch points of T' as a covering of S'. Show that $r + \delta = (4)(4)$. (One factor 4 is the number of sheets of the covering and the other is the number of boundary curves on S.)

Section 3.3

1. Dissect the locus of $zw^2 - z + 1 = 0$ into two sheets, each covering the Riemann sphere of points z. How many points on the locus cover 0, 1, ∞? How many points on the Riemann surface cover 0, 1, ∞? Give topological descriptions of the locus and the Riemann surface.

2. The roots of the equation $w^4 - 4w^2 + 4z^2 = 0$ are $w = \pm\sqrt{z + 1} \pm \sqrt{1 - z}$ (four combinations of signs). For what values of z are there

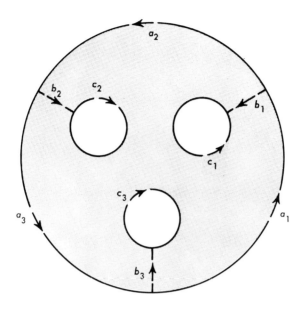

Fig. 3.13

more covering points on the Riemann surface than on the locus of the equation? Give a topological description of the Riemann surface.

3. The cubic equation $w^3 + a_1 w + a_0 = 0$ has a multiple root if and only if $4a_1^3 + 27a_0^2 = 0$. What points z are covered by fewer than three points on the locus of $zw^3 - 3zw + 2z + 1 = 0$? For what values of z is the point (z, ∞) on the locus? What points cover $z = \infty$? What is the topological nature of the Riemann surface of this equation?

4. For each non-zero value of z the four roots of the equation $w^4 z^2 - 4w^2 + 4 = 0$ may be expressed by the formula

$$w = \frac{\pm \sqrt{1 + z} \pm \sqrt{1 - z}}{z} \qquad \text{(four combinations of signs)}.$$

Find the branch points on the Riemann surface of this equation. Describe the Riemann surface topologically.

4 MAPPINGS INTO THE SPHERE

4.1 Winding Number of a Plane Curve

In this chapter we shall study mappings from a surface into the sphere. The concept of winding numbers is used to derive properties of these mappings. For concreteness we represent the sphere as the complex plane with a point at infinity adjoined. The finite points are specified either by complex numbers $z = x + iy$ or by the corresponding coordinate pairs (x, y), where x and y are Cartesian coordinates in a Euclidean plane.

In analytic geometry we use a pair of parametric equations

$$x = f(t), \qquad y = g(t),$$

to express the rectangular coordinates of the points of the curve as continuous functions of a parameter t which ranges over a real interval $[a, b]$ on which $a \le t \le b$. Using complex numbers to represent points, we can replace the two equations by a single equation

$$z = h(t)$$

where $h(t) = f(t) + ig(t)$. A third description of the curve is given by a pair of functions $r(t)$ and $\theta(t)$ which express polar coordinates of points on the curve in terms of t.

Polar coordinates present two problems: the angular coordinate is not defined at the origin, and elsewhere the angular coordinate has many values which differ by integral multiples of 2π. This means that a particular function $\theta(t)$ which represents the angular coordinate of points on a curve is undefined when $r(t) = 0$ and may be discontinuous elsewhere. For example, $\theta(t)$ could have an arbitrary number of jump discontinuities at which the height of the jump is an integral multiple of 2π. We show that every curve C not passing through the origin can be represented by functions $r(t)$ and $\theta(t)$ for which $\theta(t)$ as well as $r(t)$ is continuous. From this function $\theta(t)$ the variation $V(C)$ of the polar angle is defined as

$$V(C) = \theta(b) - \theta(a).$$

In the language of complex analysis $V(C)$ is called the *variation of the argument of z over the curve C*. If $\theta^*(t)$ were another such continuous function, $\theta(t) - \theta^*(t)$ would be a continuous function whose only values could be integral multiples of 2π. Hence $\theta(t) - \theta^*(t)$ is a constant function and $\theta^*(t) = \theta(t) + k2\pi$ for some fixed integer k. Because

$$\theta^*(b) - \theta^*(a) = \theta(b) - \theta(a),$$

we see that $V(C)$ does not depend on the selection of $\theta(t)$. A curve is closed if $h(a) = h(b)$. In terms of polar coordinates this means that

$$r(b) = r(a) \quad \text{and} \quad \theta(b) = \theta(a) + n2\pi$$

for some integer n. This integer $n = (1/2\pi)V(C)$ is called the *winding number of the curve C about the origin*.

We shall now prove that the angular coordinate can be defined as a continuous function of the parameter. Let $z = h(t)$, $a \le t \le b$ define a curve in the complex plane not passing through the origin. The function $r(t) = |h(t)| = \sqrt{f(t)^2 + g(t)^2}$ is continuous on the closed interval $[a, b]$ and assumes only positive real values. A theorem of calculus states that such a function must always have a minimum value m which is positive. By another theorem the continuous functions $f(t)$ and $g(t)$, defined on the closed interval $[a, b]$, must be uniformly continuous. As a consequence, the interval $[a, b]$ may be subdivided by a finite number of points

$$a = t_0 < t_1 < \cdots < t_j < t_{j+1} < \cdots < t_n = b,$$

so that if t is in the subinterval $[t_j, t_{j+1}]$

$$|f(t) - f(t_j)| < \frac{m}{\sqrt{2}} \quad \text{and} \quad |g(t) - g(t_j)| < \frac{m}{\sqrt{2}}.$$

This implies that

$$|h(t) - h(t_j)| = \sqrt{(f(t) - f(t_j))^2 + (g(t) - g(t_j))^2} < m.$$

Let z_j be the point $h(t_j)$ and C_j be the portion of the curve corresponding to the subinterval $[t_j, t_{j+1}]$. The last inequality says that C_j is inside the circle K_j with radius m and center z_j (Figure 4.1). Because m is the shortest

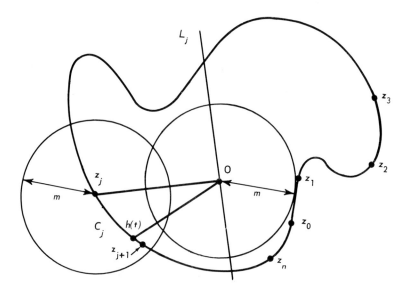

Fig. 4.1

distance from 0 to the curve, the line L_j through O perpendicular to z_j considered as a vector is either tangent to or outside of the circle K_j. The angle from z_j to $h(t)$ considered as a vector can be taken as $\psi_j(t)$, where $|\psi_j(t)| < \pi/2$. This inequality can be satisfied because C_j is inside K_j and the interior of K_j is entirely on one side of L_j. The law of cosines from trigonometry tells us that

$$|\psi_j(t)| = \arccos\left(\frac{|z_j|^2 + |h(t)|^2 - |z_j - h(t)|^2}{2|z_j||h(t)|}\right).$$

The continuity of the arc cosine function implies the continuity of $|\psi_j(t)|$ over the interval $[t_j, t_{j+1}]$. The angle $\psi_j(t)$ is positive if $h(t)$ is on one side of the line through 0 and z_j and is negative on the other side. The angle $\psi_j(t)$ can change from positive to negative only if $h(t)$ crosses the line and $\psi_j(t)$ becomes 0. Therefore the continuity of $\psi_j(t)$ now follows from the continuity of $h(t)$ and of $|\psi_j(t)|$. Let $\phi_j = \psi_j(t_{j+1})$ and let θ_0 be any angular coordinate of z_0. If t is in the interval $[t_k, t_{k+1}]$, define

$$\theta(t) = \psi_k(t) + \theta_0 + \sum_{j=0}^{k-1} \phi_j.$$

The function $\theta(t)$ defined in this way for the entire interval $[a, b]$ gives an angular coordinate for $h(t)$ which varies continuously with t (Figure 4.2).

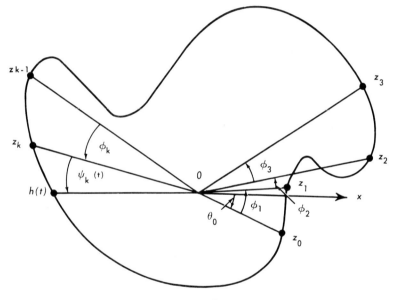

Fig. 4.2

This shows that the variation $V(C)$ is defined for every curve C. The value of $V(C)$ may be calculated from the formula

$$V(C) = \sum_{j=1}^{n} \phi_j.$$

In our discussion a curve has been specified in terms of a particular parameter. We now consider when two representations describe the same curve but with different parameters. If an equation $z = h(t)$ with $a \leq t \leq b$ describes a curve C, the order or real numbers in the interval $[a, b]$ induces an order of occurrences of points on the curve. Note that the occurrences of points and not the points themselves are ordered. This distinction is necessary because a curve may pass through the same point many times. If we think of a curve as an ordered set of occurrences of points, there are many possible parameters to describe the curve. If the same curve is specified by the equation $z = h'(s)$, where $c \leq s \leq d$, let $s = d(t)$ be the value of s such that t and $d(t)$ correspond to the same point occurrence on C. The mapping $t \to d(t)$ is a one-to-one correspondence of the interval $[a, b]$ onto the interval $[c, d]$. The mapping preserves order; that is $t < t'$ implies $d(t) < d(t')$. Because $V(C) = \sum_{j=1}^{n} \phi_j$, where ϕ_j depends only on the points z_j and not on the antecedent parameter values t_j, the variation $V(C)$ depends only on the curve and not on the selection of parameter.

If a curve C is defined by $z = h(t)$ for $a \leq t \leq b$, the restriction of $h(t)$ to two subintervals $[a, t_0]$ and $[t_0, b]$ subdivides C into two curves C_1 and C_2. Now

$$V(C) = \theta(b) - \theta(a) = \theta(b) - \theta(t_0) + \theta(t_0) - \theta(a) = V(C_1) + V(C_2).$$

Thus the angular variation over any curve may be computed by subdividing the curve and adding the variations over the segments. If the order of point occurrences on a curve C is reversed, the inverse curve C^{-1} is defined. Clearly $V(C^{-1}) = -V(C)$.

The *winding number* $\omega(C, c)$ of a closed curve C defined by $z = h(t)$ about a point $c = a + bi$ is defined to be the winding number about the origin of the curve defined by $z^* = h(t) - c$.

Example 1. We shall calculate the winding number about the point $(1, 0)$ of the curve C defined in the Euclidean plane by the parametric equations

$$\begin{aligned} x &= \cos t + \cos 2t, \\ y &= \sin t + \sin 2t, \end{aligned} \qquad 0 \leq t \leq 2\pi.$$

By definition the winding number of C about $(1, 0)$ is the same as the winding number about the origin of the curve C' defined by

$$\begin{aligned} x &= \cos t + \cos 2t - 1, \\ y &= \sin t + \sin 2t, \end{aligned} \qquad 0 \leq t \leq 2\pi.$$

We start by drawing a rough sketch (Figure 4.3) of the graphs of x and y (on C') as functions of t. We divide the t-axis into subintervals such that the curve stays within one quadrant when t stays in a subinterval. The values of t for which $x = 0$ or $y = 0$ specify endpoints of the subintervals.

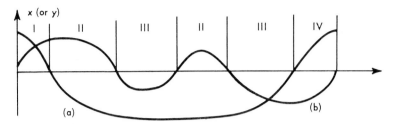

Fig. 4.3 (a) $x = \cos t + \cos 2t - 1$, (b) $y = \sin t + \sin 2t$

For C' the endpoints are $t = 0$, $2\pi/3$, π, $4\pi/3$, 2π (where $y = 0$), and

$$t = \text{arc cos}\left(\frac{-1 + \sqrt{17}}{4}\right), \quad 2\pi - \text{arc cos}\left(\frac{-1 + \sqrt{17}}{4}\right),$$

where $x = 0$. From the quadrant labels above the subintervals in Figure 4.3 we see that C' goes around the origin once. Hence the winding number of C about $(1, 0)$ is 1.

Example 2. Let the continuous function $z(t)$, defined for $0 \le t \le 1$ with $z(0) = z(1)$, determine in the complex plane a closed curve C that does not pass through the origin. We shall prove that if the directions from 0 to $z(t)$ and from 0 to $z(t + \frac{1}{2})$ are (a) always the same, then $\omega(C, 0)$ is even; (b) never the same, then $\omega(C, 0)$ is odd.

Let C_1 and C_2 be the arcs of C for which t is restricted to the intervals $I_1 : 0 \le t \le \frac{1}{2}$ and $I_2 : \frac{1}{2} \le t \le 1$, respectively. The curve C_2 may be defined by the function $z(t + \frac{1}{2})$, with I_1 as domain. Let $\theta(t)$ be a continuous definition of arg $z(t)$ for t in I_1. If the direction from 0 to $z(t + \frac{1}{2})$ is always the same as that from 0 to $z(t)$, $\theta(t)$ is also a continuous branch of arg $z(t + \frac{1}{2})$. Hence

$$V(C) = V(C_1) + V(C_2) = [\theta(\tfrac{1}{2}) - \theta(0)] + [\theta(\tfrac{1}{2}) - \theta(0)].$$

Because $z(0)$ and $z(\frac{1}{2})$ have the same ray from the origin,

$$\theta(\tfrac{1}{2}) - \theta(0) = 2k\pi$$

for some integer k. Thus

$$\omega(C, 0) = \frac{1}{2\pi} V(C) = \frac{1}{2\pi}(4k\pi) = 2k.$$

If the direction from 0 to $z(t + \frac{1}{2})$ is never the same as the direction from 0 to $z(t)$, $[z(t + \frac{1}{2})]/z(t)$ is never a positive real number, and we may define a continuous branch $\phi(t)$ of $\arg[z(t + \tfrac{1}{2})/z(t)]$ with $0 < \phi(t) < 2\pi$. The equation

$$z(t + \tfrac{1}{2}) = z(t)\,\frac{z(t + \tfrac{1}{2})}{z(t)}$$

shows that

$$\psi(t) = \theta(t) + \phi(t)$$

is a continuous definition of $\arg z(t + \tfrac{1}{2})$ for t in I_1. Now

$$V(C_2) = \psi(\tfrac{1}{2}) - \psi(0) = \theta(\tfrac{1}{2}) - \theta(0) + \phi(\tfrac{1}{2}) - \phi(0)$$
$$= V(C_1) + \phi(\tfrac{1}{2}) - \phi(0).$$

Hence

$$V(C) = V(C_1) + V(C_2) = 2V(C_1) + \phi(\tfrac{1}{2}) - \phi(0).$$

Because $V(C_1) = \theta(\tfrac{1}{2}) - \theta(0)$ is a value of

$$\arg z(\tfrac{1}{2}) - \arg z(0) = \arg \frac{z(\tfrac{1}{2})}{z(0)},$$

it follows that

$$V(C_1) = \phi(0) + 2k\pi$$

for some integer k. Thus

$$V(C) = 4k\pi + \phi(\tfrac{1}{2}) + \phi(0).$$

Now $V(C)$ is an integral multiple of 2π and $0 < \phi(\tfrac{1}{2}) + \phi(0) < 4\pi$. Hence $\phi(\tfrac{1}{2}) + \phi(0) = 2\pi$ and

$$\omega(C, 0) = \frac{1}{2\pi}\, V(C) = 2k + 1.$$

4.2   Mappings into the Plane

We consider first a continuous mapping f of an orientable surface S into the complex plane such that no point is mapped into the origin. If C is a curve on S, the ordering of occurrences of points P on C determines an ordered set of occurrences of points $f(P)$. This ordered set describes a curve denoted by $f(C)$. The variation $V(f(C))$ of this plane curve is called the variation $V_C(f(P))$ of the argument of the function f over the curve C. Let f and g be two complex valued functions, and let $\arg f(P)$ and $\arg g(P)$

be continuous branches of the argument of $f(P)$ and $g(P)$ along a curve C. The equation

$$\arg(f(P)\,g(P)) = \arg f(P) + \arg g(P)$$

defines a continuous branch of the argument of $f(P)\,g(P)$. Hence

$$V_C(f(P)\,g(P)) = V_C(f(P)) + V_C(g(P)).$$

Theorem. If S is an orientable surface, represented by the equation $a_1 b_1 a_1^{-1} b_1^{-1} \cdots a_p b_p a_p^{-1} b_p^{-1} d_1 e_1 d_1^{-1} \cdots d_r e_r d_r^{-1} = 1$, and f is a nonvanishing, continuous, complex-valued function defined on S, then

$$\sum_{i=1}^{r} V(f(e_i)) = 0.$$

Represent S as a polygon in the plane with identified edges such that this equation describes the boundary in a counterclockwise direction. With this model for S, f is a continuous function defined on the polygon subject to the condition that f has the same values at identified points. The continuity of f on a closed bounded set implies that f is uniformly continuous and that $|f(P)|$ has a minimum m on S. Uniform continuity ensures that we can subdivide S into a finite number of polygons π_j which are small enough so that $|f(P) - f(P')| < m$ for any two points in the same polygon. If Γ_j is the boundary of a polygon of the subdivision, the closed curve $f(\Gamma_j)$ is within a circle of radius m with its center at any point of the curve (Figure 4.4). Therefore $V(f(\Gamma_j)) = 0$. Suppose that each curve Γ_j is oriented counterclockwise. Now

$$\sum_{j} V(f(\Gamma_j)) = 0.$$

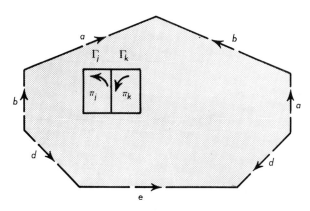

Fig. 4.4

The sum $\sum_j V(f(\Gamma_j))$ can be rewritten as the sum of the variations of the edges which make up the curves Γ_j. Each edge x that is not a segment of an edge of S is on the boundary of two of the π_j. In one of these boundary curves the edge is oriented clockwise and in the other, counterclockwise. This means that $V(f(x))$ appears once positively and once negatively. The sum $\sum_j V(f(\Gamma_j))$ reduces to the sum of the variations over the edges of S. Because each edge a_i, b_i, or d_i occurs once in each orientation, we can conclude that

$$\sum_{i=1}^{r} V(f(e_i)) = 0.$$

Brouwer Fixed-Point Theorem (two-dimensional case). Any continuous mapping of a disk into itself has at least one fixed point.

Represent the disk by the locus in the complex plane of the inequality $|z| \leq 1$. The mapping becomes a continuous complex function f defined on the disk with the property that $|f(z)| \leq 1$ for all z. On the circle C, defined by $|z| = 1$, $|f(z)/z| \leq 1$. Now

$$\left| \text{Re}\left(\frac{f(z)}{z} \right) \right| \leq \sqrt{ \left\{ \text{Re}\left[\frac{f(z)}{z} \right] \right\}^2 + \left\{ \text{Im}\left[\frac{f(z)}{z} \right] \right\}^2 } = \left| \frac{f(z)}{z} \right|.$$

Therefore on C

$$\text{Re}\left[\frac{f(z)}{z} - 1 \right] = \text{Re}\left[\frac{f(z)}{z} \right] - 1 \leq \left| \text{Re}\left[\frac{f(z)}{z} \right] \right| - 1 \leq 0.$$

If $\text{Re}[f(z)/z - 1] = 0$, $\text{Re}[f(z)/z] = 1$ and

$$\sqrt{ 1 + \text{Im}\left[\frac{f(z)}{z} \right]^2 } \leq 1.$$

Hence

$$\text{Im}\left(\frac{f(z)}{z} \right) = 0, \quad \frac{f(z)}{z} = 1, \quad \text{and} \quad f(z) = z.$$

If this happens, z is a fixed point and the theorem is proved. If $f(z) \neq z$ for any point on C, $\text{Re}[f(z)/z - 1] < 0$ for all points on C. Therefore

$$V_C\left[\frac{f(z)}{z} - 1 \right] = 0.$$

Now

$$V_C[f(z) - z] = V_C(z) + V_C[f(z)/z - 1] = V_C(z).$$

If the counterclockwise orientation is assigned to C, $V_C(z) = 2\pi$; but the disk is an orientable surface with the single boundary curve C, and so the previous theorem requires that

$$V_C[f(z) - z] = 0,$$

unless $f(z) - z = 0$ for some point z in the disk. Hence there must be at least one fixed point z at which

$$f(z) = z.$$

Borsuk-Ulam Theorem (two-dimensional case). A continuous mapping of a Euclidean sphere into the plane must map some pair of antipodal points onto a single point.

Let P and P^* be antipodal points on the sphere S. If f is the mapping, consider the new function $g(P) = f(P) - f(P^*)$. Let C be a great circle on the sphere and let H be one of the hemispheres bounded by C. If there is no point P on S such that $f(P) = f(P^*)$, g defines a mapping of H into the complex plane such that $g(P)$ is never 0. Since H has the single boundary curve C, $\omega(g(C), 0) = (1/2\pi)V(g(C)) = 0$. On the other hand, we shall show that the relation $g(P) = -g(P^*)$ implies that $\omega(g(C), 0)$ is odd. This will prove that there is at least one point P such that $f(P) = f(P^*)$.

Divide C into two semicircles C_1 and C_2 with endpoints P_1 and P_1^*. Since $g(P_1) = -g(P_1^*)$,

$$V(g(C_1)) = \pi + n2\pi$$

for some integer n. The formula

$$\arg(g(P^*)) = \pi + \arg(g(P))$$

defines a continuous branch of the argument of $g(P^*)$ for P^* on C_2 from a given continuous branch of the argument of $g(P)$ on C_1. Hence

$$V(g(C_2)) = V(g(C_1)) \quad \text{and} \quad \omega(g(C), 0) = (1/2\pi)V(g(C)) = 1 + 2n.$$

The ham sandwich theorem is a corollary of the Borsuk-Ulam theorem. A ham sandwich consists of three ingredients: the ham, the butter, and the bread, filling three volumes in Euclidean three-space. The sandwich will be divided fairly between two people if each person receives half of each constituent. The problem is to bisect all three ingredients with a single slice of a knife.

Ham Sandwich Theorem. For any three solids S_1, S_2, S_3 (with finite volumes) in Euclidean three-space there is a plane that simultaneously bisects all three solids.

Among the planes perpendicular to a given unit vector there are three π_1, π_2, and π_3, which bisect S_1, S_2, and S_3, respectively. Let x be the directed distance from π_1 to π_2 and y from π_1 to π_3. These distances will be positive if measured in the direction of the unit vector and negative if measured in the opposite direction. If we identify the unit vector with the corresponding point of the unit sphere, the mapping $P \rightarrow (x, y)$ is a continuous transformation of the sphere into the plane. If P^* is the antipode of P, the planes π_1, π_2, π_3 are the same for P^* and P but the direction for measuring positive distance is reversed. Hence $f(P) = -f(P^*)$; but by the Borsuk-Ulam Theorem there is a point P_0 such that $f(P_0) = f(P_0^*)$. For this point $f(P_0) = -f(P_0)$, hence $f(P_0) = (0, 0)$. The three planes corresponding to P_0 coincide; that is, there is a single plane bisecting all three solids.

4.3   The Brouwer Degree

We have considered mappings into the plane such that 0 is not an image. We now discuss mappings for which a finite number of points are mapped onto 0. First we must learn to count these roots, or zeros, with appropriate multiplicities.

Consider a continuous mapping f from an orientable surface S into the plane. Assume that $f(P) = 0$ only at a finite set of points: $P_1, \ldots, P_k$, none of which is on a boundary curve of S. In S select nonintersecting disks $\Delta_1, \ldots, \Delta_k$ which contain $P_1, \ldots, P_k$, respectively as interior points. A canonical equation for the surface has the form

$$a_1 b_1 a_1^{-1} b_1^{-1} \cdots a_p b_p a_p^{-1} b_p^{-1} d_1 e_1 d_1^{-1} \cdots d_r e_r d_r^{-1} = 1.$$

We may select the edges in this equation so that none intersects the disks.

We define an orientation on S by agreeing that the closed curve in the canonical equation is positively oriented. Let c be the boundary curve of a disk Δ in S. Assume that Δ intersects no edge of S. If c starts and finishes at the point P, let y be a simple curve starting at the single vertex of the canonical equation and finishing at P. Assume that no intermediate point on y is on any edge of S or in Δ. We say that c is *positively oriented* if the equation

$$a_1 b_1 a_1^{-1} b_1^{-1} \cdots a_p b_p a_p^{-1} b_p^{-1} d_1 e_1 d_1^{-1} \cdots d_r e_r d_r^{-1} y c^{-1} y^{-1} = 1$$

describes the surface derived from S by deleting the interior of Δ. Interpret the canonical equation as a counterclockwise description of the boundary of a planar polygon from which S is obtained by edge identifications.

Our definition says that the counterclockwise direction is the positive orientation of the boundary of a disk Δ inside the polygon (Figure 4.5). Notice that the curve in the canonical equation is also a disk boundary if the edge identifications are ignored. Although we shall not need to know the positive orientation on the boundary curves of disks crossing the edges of the polygon, the orientability of S would permit the definition of positive orientation to be extended to all disk boundaries in S. No satisfactory extension could be made if S were nonorientable.

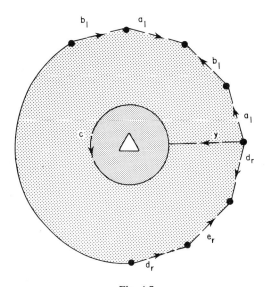

Fig. 4.5

We now apply our orientation to the boundary curves of the disks $\Delta_1, \ldots, \Delta_k$. Let $Q_1, \ldots, Q_k$ be points on the boundary curves of the disks $\Delta_1, \ldots, \Delta_k$, respectively, and let $c_1, \ldots, c_k$ be the positively oriented boundary curves of the disks starting and finishing at $Q_1, \ldots, Q_k$. Let $y_1, \ldots, y_k$ be edges from the vertex of S to $Q_1, \ldots, Q_k$, respectively. Assume that no y_j intersects any other labeled edge or curve except at the endpoints of y_j. The surface S' derived from S by deleting the interior of each disk has the canonical equation

$$a_1 b_1 a_1^{-1} b_1^{-1} \cdots a_p b_p a_p^{-1} b_p^{-1} d_1 e_1 d_1^{-1} \cdots d_r e_r d_r^{-1} y_1 c_1^{-1} y_1^{-1} \cdots y_k c_k^{-1} y_k^{-1} = 1.$$

Because $f(P)$ is never zero for P in S',

$$\sum_{i=1}^{r} V(f(e_i)) + \sum_{j=1}^{k} V(f(c_j^{-1})) = 0.$$

Equivalently,

$$\sum_{i=1}^{r} V(f(e_i)) = \sum_{j=1}^{k} V(f(c_j)).$$

Because the e_i and c_j are closed curves, this equation may be replaced by the following equation in integers:

(1) $$\sum_{i=1}^{r} \omega(f(e_i), 0) = \sum_{j=1}^{k} \omega(f(c_j), 0).$$

The winding number $\omega(f(c_j), 0)$ is called the *multiplicity* or *order* $\mu(P_j, 0)$ of P_j as a zero of f.

We now show that $\mu(P_j, 0)$ does not depend on the selection of Δ_j. If D is a second disk which determines $\mu(P_j, 0)$, let D' be a disk with P_j in its interior such that $D' \subset \Delta_j \cap D$. By showing that Δ_j and D' determine the same value of $\mu(P_j, 0)$ and that D and D' also specify equal values of $\mu(P_j, 0)$ we will have shown that $\mu(P_j, 0)$ does not depend on Δ_j. Let c_j' be the positively oriented boundary curve of D'. The function f defined on Δ_j has a zero only at P_j which is inside D'. Because the disk Δ_j is an orientable surface with a single boundary curve c_j, application of (1) to Δ_j gives

$$\omega(f(c_j), 0) = \omega(f(c_j'), 0).$$

Thus Δ_j and D' determine the same value of $\mu(P_j, 0)$. The same reasoning applies to D and D'.

We have found that the multiplicity or order of a zero is uniquely determined as soon as a positive orientation is specified. All multiplicities would by multiplied by -1 if the orientation were reversed. Because a reversal would also change the orientation of the boundary curves e_i in the canonical equation, the formula

(1a) $$\sum_{i=1}^{r} \omega(f(e_i), 0) = \sum_{j=1}^{k} \mu(P_j, 0)$$

would still be valid.

Consider mappings of an orientable surface S into the sphere or extended complex plane. A continuous mapping f is now allowed to have $f(Q) = \infty$ at a finite set of points, $Q_1, \ldots, Q_n$. If we think of 0 and ∞ as the north and south poles of a Euclidean sphere, a parallel of latitude would go around each pole once, but the orientation, which is clockwise, to an observer at one pole is counterclockwise to an observer at the other. This suggests the definition

$$\omega(C, \infty) = -\omega(C, 0)$$

for the winding number $\omega(C, \infty)$ of a curve C about ∞. If the mapping f has $f(Q) = \infty$, we say f has a *pole* at Q and the *multiplicity* or *order* of the pole is

$$\mu(Q, \infty) = \omega(f(x), \infty) = -\omega(f(x), 0),$$

where x is the positively oriented boundary of a sufficiently small disk about Q. We can now derive the equation

(2)
$$\sum_{i=1}^{r} \omega(f(e_i), 0) = \sum_{j=1}^{k} \mu(P_j, 0) - \sum_{j=1}^{n} \mu(Q_j, \infty),$$

where P_j are the zeros and Q_j the poles of f. This formula says that for a function with only a finite number of zeros and poles the number of zeros minus the number of poles counted with their multiplicities depends only on the behavior of f on the boundary curves of S.

We define $\mu(P, 0) = 0$ if $f(P) \neq 0$ and $\mu(P, \infty) = 0$ if $f(P) \neq \infty$. Formula (2) may be rewritten as

$$\sum_{i=1}^{r} \omega(f(e_i), 0) = \sum_{P \in S} \mu(P, 0) - \sum_{P \in S} \mu(P, \infty).$$

These infinite sums exist because there is only a finite number of nonzero terms. This formula has special significance when S is a closed surface. When there are no curves e_i,

$$\sum_{P \in S} \mu(P, 0) = \sum_{P \in S} \mu(P, \infty).$$

If $f(P) = z_0$, the multiplicity $\mu(P, z_0)$ with which f takes the value z_0 at P is the order of the zero at P of the function h defined by $h(P) = f(P) - z_0$. Now h has the same poles as f. If a point is near enough to a pole, $|f(P)|$ may be assumed to be larger than any fixed real constant and, in particular, much larger than $|z_0|$. This means that the additive constant z_0 cannot change the winding number of $f(P)$ as P traverses a small curve around a pole of f. Hence f and h have the same poles with the same multiplicities. Thus

$$\sum_{P \in S} \mu_f(P, z_0) = \sum_{P \in S} \mu_h(P, 0) = \sum_{P \in S} \mu_h(P, \infty) = \sum_{P \in S} \mu_f(P, \infty).$$

We have now proved the following theorem.

Theorem. If f is a continuous mapping of a closed orientable surface into a sphere such that no value is taken an infinite number of times, every point on the sphere appears as an image the same number of times, provided that these appearances are counted with their multiplicities.

The number of times each value is assumed is called the *Brouwer degree* of the mapping. If the positive orientation of S is reversed, the Brouwer degree is multiplied by -1.

The projection of a Riemann surface S' onto the sphere S is the mapping that projects every point on the covering surface onto the point covered. A small disk Δ on S is covered by a set of disks in S'. The positive orientation of the boundary of a covering disk Δ' is defined as the orientation induced by the positive direction on the boundary of Δ. Let P be a point of Δ and P', the point of Δ' covering P. The point P' will be unique if Δ is sufficiently small. When Δ is this small, no point of Δ' except possibly P' can be a branch point of P. The number of times the projection of the boundary of Δ' covers the boundary of Δ is the multiplicity with which P' covers P. This multiplicity is one unless P' is a branch point. The order of a branch point is one less than its multiplicity. The Brouwer degree of the projection of a Riemann surface onto the sphere is the number of sheets in the covering.

4.4 Applications of the Winding Number in Complex Analysis

On the sphere represented as the extended complex plane we define the positive orientation of the boundary of a disk as counterclockwise if the disk is bounded and clockwise if it has ∞ as an interior point.

We shall study the mappings of a sphere S into itself defined by rational functions of the form $f(z) = p(z)/q(z)$, where $p(z) = a_0 + \cdots + a_k z^k$ and $q(z) = b_0 + \cdots + b_n z^n$ with $a_k \neq 0$ and $b_n \neq 0$. We assume that the polynomials $p(z)$ and $q(z)$ have no common factors other than constants. This means that the fraction $p(z)/q(z)$ has been reduced to its lowest form. The function $f(z)$ may be rewritten in the form

$$f(z) = (z - z_1)^{e_1} \frac{p_1(z)}{q_1(z)},$$

where $(z - z_1)$ is not a factor of $p_1(z)$ or $q_1(z)$. If e_1 is a positive integer, $p(z) = (z - z_1)^{e_1} p_1(z)$ and $q(z) = q_1(z)$. Thus z_1 is a zero of $p(z)$ and $f(z)$. If e_1 is a negative integer, $q(z) = (z - z_1)^{-e_1} q_1(z)$ and $p(z) = p_1(z)$ so that z_1 is a zero of $q(z)$ and a pole of $f(z)$. When $e_1 = 0$, $p(z) = p_1(z)$ and

$q(z) = q_1(z)$. In this case z_1 is neither a zero nor a pole of $f(z)$. The behavior of $f(z)$ when $z = \infty$ may be studied by expressing $f(z)$ in the form

$$f(z) = \left(\frac{1}{z}\right)^{n-k} \frac{p_\infty(1/z)}{q_\infty(1/z)},$$

where

$$p_\infty\left(\frac{1}{z}\right) = \frac{a_0}{z^k} + \cdots + \frac{a_{k-1}}{z} + a_k$$

and

$$q_\infty\left(\frac{1}{z}\right) = \frac{b_0}{z^n} + \cdots + \frac{b_{n-1}}{z} + b_n.$$

The function $f(z)$ has a zero or pole at ∞ when $n > k$ or $n < k$, respectively. If $n = k$, $f(\infty) = a_k/b_n$ and $f(z)$ has neither a zero nor a pole at ∞.

By elementary algebra we know that $p(z)$ and $q(z)$ have at most as many zeros as their degree. Therefore the number of zeros and poles of $f(z)$ is finite. The multiplicity of a zero z_1 of $f(z)$ is

$$\mu(z_1, 0) - \frac{1}{2\pi} V_C(f(z)),$$

where C is the positively oriented boundary of a small disk with z_1 in its interior. Now

$$V_C(f(z)) = V_C\left((z - z_1)^{e_1} \frac{p_1(z)}{q_1(z)}\right) = e_1 V_C(z - z_1) + V_C\left(\frac{p_1(z)}{q_1(z)}\right).$$

Let

$$r = \left|\frac{q_1(z_1)}{q_1(z_1)}\right|.$$

The disk bounded by C can be selected so small that $p_1(z)/q_1(z)$ for all z on C stays inside a circle of radius r and center $p_1(z_1)/q_1(z_1)$. For such a disk $V_C(p_1(z)/q_1(z)) = 0$. Because $V_C(z - z_1) = 2\pi$, we conclude that $\mu(z_1, 0) = e_1$. If z_1 had been a pole of $f(z)$ instead of a zero, the formula $V_C(f(z)) = 2\pi e_1$ would still be valid. In this case $\mu(z_1, \infty) = -e_1 > 0$. We have found that our topological definition of the multiplicity of a zero or pole agrees with the familiar algebraic definition of the order of a zero or pole of a rational function.

We shall now study zeros or poles at $z_1 = \infty$. Let C be a circle large enough so that all the finite complex numbers that are zeros or poles of $f(z)$ are inside C. The region outside C, together with C, forms a disk containing ∞ and no other zero or pole of $f(z)$. The positive orientation of C as the boundary of the disk is clockwise. Now

$$V_C(f(z)) = V_C\left(\left(\frac{1}{z}\right)^{n-k}\frac{p_\infty(1/z)}{q_\infty(1/z)}\right) = (n-k)\,V_C\left(\frac{1}{z}\right) + V_C\left(\frac{p_\infty(1/z)}{q_\infty(1/z)}\right).$$

If C is large enough $p_\infty(1/z)/q_\infty(1/z)$ stays close to a_k/b_n so that

$$V_C\left(\frac{p_\infty(1/z)}{q_\infty(1/z)}\right) = 0.$$

Because $V_C(1/z) = -V_C(z) = -(-2\pi) = 2\pi$, we find that $\mu(\infty, 0) = n - k$ if $n > k$ or $\mu(\infty, \infty) = k - n$ if $n < k$.

We now consider the special rational functions for which $q(z)$ is the constant polynomial with value 1. Then $f(z)$ is a polynomial of degree k with $\mu(\infty, \infty) = k$. Because

$$\sum_{z \in S} \mu(z, 0) = \sum_{z \in S} \mu(z, \infty) = \mu(\infty, \infty) = k,$$

the polynomial of degree k has k zeros when the zeros are counted with their multiplicities. A consequence is the following theorem proved by Gauss:

Fundamental Theorem of Algebra. If $f(z)$ is a nonconstant polynomial with complex coefficients, there is at least one complex number z_0 such that $f(z_0) - 0$.

The system of complex numbers is an example of a *field*. One branch of algebra is concerned with fields and the roots of polynomial equations with coefficients in a field. A field F is algebraically closed if for every nonconstant polynomial $p(x)$, with coefficients in F, the equation $p(x) = 0$ has a root in F. The theorem states that the field of complex numbers, a field of special interest in analysis, is one of the many algebraically closed fields. Furthermore, the usual proofs are topological or analytic. Thus the fundamental theorem of algebra is most properly classified as a theorem of analysis.

Returning to the rational function $f(z) = p(z)/q(z)$, we now know that the number of zeros or poles in the finite complex plane equals the degree $p(z)$ or $q(z)$, respectively. When ∞ is also considered, the number of zeros and the number of poles both equal the maximum of the degree of the numerator and the degree of the denominator. This maximum is called the degree of the rational function. We have now proved the following theorem.

Theorem. The Brouwer degree of the mapping of the Riemann sphere into itself defined by a rational function equals the degree of the rational function.

We have used the winding number to prove that a polynomial equation has complex roots. We shall now see that the winding number can help to isolate these roots. Let f be a function mapping a surface S into the extended complex plane. When the winding number of f around the boundary of a disk D in S is not zero, there is at least one zero or pole in D. If we subdivide D into two disks, D_1 and D_2, so that there is no zero or pole on any boundary curve, the winding number of f over the boundary of D is the sum of the winding numbers of f over the boundaries of D_1 and D_2. Hence the winding number of f is nonzero over the boundary of at least one of the disks D_1 and D_2. By repeated subdivision of D, we can locate a zero or pole of f in smaller and smaller regions. If S is embedded in Euclidean three-space, the zero or pole may be located inside a sphere of radius ε for any prescribed positive ε. We shall illustrate this procedure by isolating the roots of a fourth-degree polynomial.

First we shall prove Rouche's theorem because it is useful for calculating winding numbers.

Rouche's Theorem. If $f(P)$ and $g(P)$ are two complex valued functions defined over a closed curve C and $|f(P)| < |g(P)| < \infty$ for all P on C, then

$$V_C(f(P) + g(P)) = V_C(g(P)).$$

Because $|f(P)| < |g(P)|$, $g(P) \neq 0$ for any P on C. Now

$$V_C(f(P) + g(P)) = V_C(g(P)) + V_C\left(1 + \frac{f(P)}{g(P)}\right).$$

Because $|f(P)/g(P)| < 1$, $1 + f(P)/g(P)$ stays inside the circle $|z - 1| = 1$. Because $1 + f(P)/g(P)$ is never in the second or third quadrant,

$$V_C(1 + f(P)/g(P)) = 0$$

and Rouche's theorem follows.

If the hypothesis that C is closed is dropped, this reasoning shows that $|V_C(f(P) + g(P)) - V_C(g(P))| < \pi$.

Example 3. Isolate the roots of the polynomial equation

$$f(z) = z^4 + 3z^2 - 6z + 10 = 0.$$

Consider the circular disk D defined by $|z| \leq 3$. On the circle C bounding D

$$|3z^2 - 6z + 10| \leq |3z^2| + |6z| + 10 = 27 + 18 + 10 = 55 < 81 = |z^4|.$$

By Rouche's theorem

$$V_C(z^4 + 3z^2 - 6z + 10) = V_C(z^4) = 4V_C(z) = 8\pi.$$

Hence $f(z)$ has all four of its zeros inside C. Let D_1, D_2, D_3, D_4 be the portions of D and c_1, c_2, c_3, c_4 the arcs of C in the first, second, third, and fourth quadrants, respectively. (Figure 4.6). Let x_1, y_1, x_2, y_2 be the

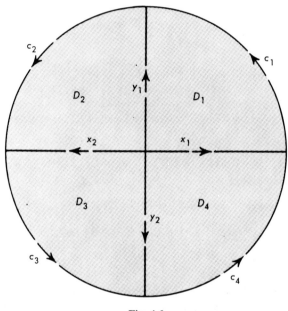

Fig. 4.6

radii of D along the positive x-axis, the positive y-axis, the negative x-axis, and the negative y-axis, respectively. The polygons D_1, D_2, D_3, D_4 are described by the edge equations

$$x_1 c_1 y_1^{-1} = 1, \qquad y_1 c_2 x_2^{-1} = 1, \qquad x_2 c_3 y_2^{-1} = 1, \qquad y_2 c_4 x_1^{-1} = 1.$$

Let B_1, B_2, B_3, B_4 be the boundary curves of D_1, D_2, D_3, D_4, respectively. Now

$$V_{B_1}(f(z)) = V_{c_1}(f(z)) + V_{x_1}(f(z)) - V_{y_1}(f(z)).$$

Similar equations hold for the variation over B_2, B_3, and B_4. By the extension of Rouche's theorem

$$V_{c_j}(f(z)) = V_{c_j}(z^4) + \varepsilon_j = 2\pi + \varepsilon_j, \qquad j = 1, 2, 3, 4,$$

where $|\varepsilon_j| < \pi$. If z is a real number x,

$$f(z) = x^4 + 3x^2 - 6x + 10 = x^4 + 3(x - 1)^2 + 7 > 0.$$

Hence $V_{x_1}(f(z)) = V_{x_2}(f(z)) = 0$. If z is an imaginary number iy,

$$f(z) = y^4 - 3y^2 + 10 - 6iy.$$

Because

$$y^4 - 3y^2 + 10 = \left(y^2 - \frac{3}{2}\right)^2 + \frac{31}{4} > 0,$$

Re $f(z) > 0$ on y_1 and y_2. Hence $V_{y_1}(f(z)) = \eta_1$ and $V_{y_2}(f(z)) = \eta_2$ where $|\eta_1| < \pi$ and $|\eta_2| < \pi$. Therefore

$$\omega_{B_1}(f(z), 0) = \frac{1}{2\pi} V_{B_1}(f(z)) = 1 + \frac{1}{2\pi}(\varepsilon_1 - \eta_1),$$

$$\omega_{B_2}(f(z), 0) = 1 + \frac{(\varepsilon_2 + \eta_1)}{2\pi},$$

$$\omega_{B_3}(f(z), 0) = 1 + \frac{(\varepsilon_3 - \eta_2)}{2\pi},$$

$$\omega_{B_4}(f(z), 0) = 1 + \frac{(\varepsilon_4 + \eta_2)}{2\pi}.$$

Because the winding numbers are integers and $|\varepsilon_j/2\pi| < \frac{1}{2}$ and $|\eta_j/2\pi| < \frac{1}{2}$ for each j, $\varepsilon_1 = \eta_1 = -\varepsilon_2$ and $\varepsilon_3 = \eta_2 = -\varepsilon_4$. Thus each winding number is 1, and there is one root of $f(z) = 0$ inside each quadrant of the disk $|z| \leq 3$. We have now isolated the four roots in separate regions.

EXERCISES

Section 4.1

1. The equation

$$z(t) = 2\cos t + 2i\sin t + 1, \qquad 0 \leq t \leq 2\pi,$$

defines a circle with center at $z = 1$. What is the minimum value m of $|z(t)|$? Find a set of numbers

$$t_0 = 0 < t_1 < t_2 < \cdots < t_j < \cdots < t_{k-1} < t_k = 2\pi$$

such that $|z(t) - z(t_j)| < m$ if $t_j \leq t \leq t_{j+1}$.

2. Calculate the winding number about the points $(-\frac{1}{2}, 0)$ and $(-2, 0)$ of the curve defined by the parametric equations

$$\begin{aligned} x &= \cos t + \cos 2t, \\ y &= \sin t + \sin 2t, \end{aligned} \qquad 0 \leq t \leq 2\pi.$$

WARNING. Do not try to deduce the answers from Figure 4.3. The graphs of the figure are not sufficiently accurate.

3. Find the winding number about the origin of the curves defined by the following:

(a) $z(t) = \cos nt + i \sin nt,$ $0 \leq t \leq 2\pi,$
(b) $z(t) = \cos nt - i \sin nt,$ $0 \leq t \leq 2\pi.$

4. Let the continuous function $z(t)$ defined for $0 \leq t \leq 1$ with $z(0) = z(1)$ determine a closed curve C not passing through 0.

(a) Show that the winding number of C about 0 is even if the direction from 0 to $z(t)$ is never opposite to the direction from 0 to $z(t + \frac{1}{2})$ for $0 \leq t \leq \frac{1}{2}$.
(b) Show that the winding number of C about 0 is odd if the direction from 0 to $z(t)$ is always opposite to the direction from 0 to $z(t + \frac{1}{2})$ for $0 \leq t \leq \frac{1}{2}$.

5. Let the continuous functions $z_1(t)$ and $z_2(t)$ defined for $0 \leq t \leq 1$ with $z_1(0) = z_1(1)$ and $z_2(0) = z_2(1)$ determine two closed curves that do not pass through 0. Show that the winding number about 0 is the same for both curves if the line segment from $z_1(t)$ to $z_2(t)$ does not contain 0 for any value of t.

Section 4.2

1. Consider the complex function

$$f(z) = z^2 + (|z| - 1)z,$$

defined in the annulus $1 \leq |z| \leq 2$. Determine $V_C(f(z))$, where C is the circle $|z| = 2$ with the counterclockwise orientation.

2. Let $f(z)$ be a continuous complex function defined on the disk $|z| \leq 1$ with the property that $|f(z)| = 1$ for all z. Show that $f(z) = z$ for some value of z and $f(z) = -z$ for some other value of z.

3. Prove the pancake theorem. Two pancakes on a griddle (two regions in a plane) may be simultaneously bisected by a single stroke of a knife (by a line).

4. Prove the Brouwer fixed-point theorem (one-dimensional case). Any continuous mapping of a line segment into itself has a fixed point.

5. Prove the Borsuk-Ulam theorem (one-dimensional case). Any continuous mapping of a circle into a line maps some pair of antipodal points onto a single point.

6. Let f be a continuous mapping of the sphere S into itself. Prove that if there is a point Q in S such that $f(P)$ does not equal Q for any P in S the mapping f has a fixed point.
HINT. Apply the Brouwer fixed-point theorem to S with a small neighborhood of Q deleted.

Section 4.3

1. What are the zeros and poles of the function

$$f(z) = z + \frac{1}{z},$$

defined in the disk $|z| \leq 2$? If C is the circle $|z| = 2$ with counterclockwise orientation, calculate the winding number $\omega_C(f(z),0)$.

2. If $z = r(\cos \theta + i \sin \theta)$, let $f(z) = r(\cos 3\theta + i \sin 3\theta)$. Also define $f(\infty) = \infty$. For the mapping f of the Riemann sphere into itself find $\mu(0, 0)$, $\mu(i, -i)$, and $\mu(\infty, \infty)$. What is the Brouwer degree of f?

3. In a Euclidean three-space with rectangular coordinates x, y, z, let S be the torus defined by the equation

$$z^2 = (25 - x^2 - y^2)(x^2 + y^2 - 1).$$

Let S_0 be the sphere $x^2 + y^2 + z^2 = 1$, Q, the point $(1, 0, 0)$, and R, the point $(0, 0, 1)$.

(a) If P is a point (x, y, z) on S, define $f(P)$ to be the point $(x/r, y/r, z/r)$, where $r = \sqrt{x^2 + y^2 + z^2}$. What points P have $f(P) = Q$? For each of these points determine $\mu(P, Q)$. What points P have $f(P) = R$? What is the Brouwer degree of the mapping f of S into S_0?

(b) If P is a point (x, y, z) on S, define $g(P)$ to be the point $[(x + 3)/t,$ $y/t, z/t]$, where $t = \sqrt{(x + 3)^2 + y^2 + z^2}$. What points P satisfy $g(P) = R$? What is the Brouwer degree of the mapping g of S into S_0?

Section 4.4

1. Isolate the roots of the polynomial equation

$$z^4 - 2z^3 - z^2 + 2z + 10 = 0.$$

Note.

$$x^4 - 2x^3 - x^2 + 2x + 10 = \tfrac{1}{2}(x - 2)^2 x^2 + \tfrac{1}{2}(x^2 - 4)^2 + (x + 1)^2 + 1.$$

2. Show that the polynomial equation

$$z^3 + z^2 + z + 4 = 0$$

has one root in the first quadrant, one in the fourth quadrant, and one on the negative real axis.

3. Express in terms of a_0, a_1, and a_2 a real number R such that all roots of the equation

$$z^3 + a_2 z^2 + a_1 z + a_0 = 0$$

satisfy the inequality $|z| < R$.

4. Show that the roots of the equation

$$z^6 + 2z^3 + 3z^2 + 5z + 15 = 0$$

all lie in the annulus $1 < |z| < 2$.

5 VECTOR FIELDS

5.1 Vector Fields on the Plane

In Chapter 4 we studied mappings into the plane defined by complex valued functions. The complex number $z = x + iy$ or the coordinate pair (x, y), appearing as a value of a function, may be considered as a vector instead of a point. When $f(P)$ represents a point, the winding number $\omega_C(f(P), 0)$ counts the number of times the image of a curve C winds about the origin. When $f(P)$ is interpreted as a vector, the winding number is the number of revolutions of the direction of $f(P)$ as P travels along C. If the function f is defined over a region R of the complex plane, we say that the vectors $f(z)$ form a vector field on R. We represent the vector $f(z)$ by the oriented line segment from the point z to the point $z + f(z)$.

The vector interpretation clarifies the role of the function $h(z) = f(z) - z$ in the proof of the Brouwer fixed-point theorem. This function assigns to each point z the vector from z to its image $f(z)$. The condition that f maps the disk into itself requires that each vector on the circumference of the disk be directed into the disk. The Brouwer fixed-point theorem is equivalent to the following theorem.

Theorem. If a continuous vector field defined over a circular disk has all vectors on the circumference directed into the disk, the zero vector is assigned to some point of the disk.

A point is a *singularity* of a continuous vector field on the plane if the vector at that point is the zero vector. A singularity is *isolated* if it has a neighborhood that contains no other singularity. We define the *index* $I(z_0)$ of an isolated singularity z_0 of a vector field defined by $f(z)$ as the multiplicity of z_0 as a zero of $f(z)$.

As an application of the theorem consider the hair on a person's head. The scalp plays the part of a disk and the hair the role of the vector field. If the hair, untrained by brush or comb, is allowed to lie naturally, the hair at the edge of the scalp will point away from the top of the head. As predicted by the theorem, we find a singularity, the cowlick or hair swirl at the back of the head (Figure 5.1).

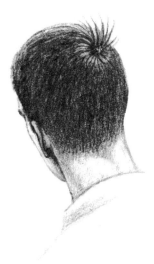

Fig. 5.1 Boy with a cowlick

A corollary of formula (1a) in Section 3, Chapter 4, is the following (Figure 5.2).

Theorem. Suppose f defines a continuous vector field over the multiply connected plane region consisting of points inside a counterclockwise

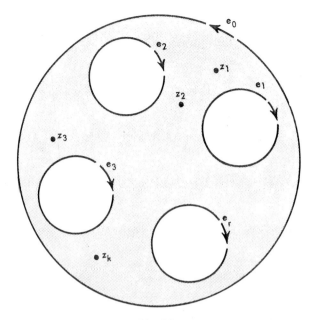

Fig. 5.2

oriented curve e_0 and outside the clockwise oriented curves $e_1, \ldots, e_r$. If the vector field has singularities only at the interior points $z_1, \ldots, z_k$, then

$$\sum_{i=0}^{r} \omega(f(e_i), 0) = \sum_{j=1}^{k} I(z_j).$$

If x and y are rectangular coordinates in a region R, let $u(x, y)$ and $v(x, y)$ be the x- and y-coordinates of the vectors of a continuous vector field over R. Curves in the xy-plane which satisfy the differential equations

$$\frac{dx}{dt} = u(x, y) \qquad \text{and} \qquad \frac{dy}{dt} = v(x, y)$$

are tangent everywhere to the vector field. The tangent vectors point in the direction of increasing values of the parameter t. A sketch of these solution curves is a graphic representation of the vector field. At a singularity of the vector field $u(x, y) = 0$ and $v(x, y) = 0$. This means that the direction of the solution curve is undefined.

Sources and *sinks* are two types of singularities. At a source all curves are directed away from the singularity, whereas at a sink the curves lead

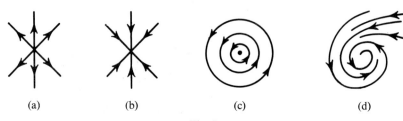

Fig. 5.3
Singularities of Index 1. (a) Source, (b) sink, (c) vortex, (d) focus

to the singularity (Figure 5.3). Because the tangent vectors to the curves complete one revolution in the positive direction as a point moves counterclockwise around a source or sink, the index is 1. *Vortices* and other *foci* are also singularities of index 1. A *dipole* results if a source and sink are brought together (Figure 5.4). Moving two vortices together can also create a dipole. The index of a dipole is 2. Sources and sinks can be combined to construct singularities with an arbitrary positive integer as index. A *simple crosspoint* or *col* is a singularity of index -1 (Figure 5.5). At a simple crosspoint two curves meet. A crosspoint of index $-m$ is formed if $m + 1$ curves meet.

Example 1. In Euclidean plane geometry consider a continuous field of nonzero vectors defined over a circular disk. We shall prove that there is at least one point on the boundary of the disk at which the vector points directly away from the center of the disk.

Let the center of the disk be the origin of coordinates and let $\theta(P)$ with $0 \leq \theta \leq 2\pi$ be the angular polar coordinate of a point P on the boundary of the disk. Let $\phi(\theta)$ be a continuous function of θ whose value is the angle from the direction of the positive x-axis to the direction of the vector at P. Define $\psi(\theta) = \phi(\theta) - \theta$. As θ increases from 0 to 2π, $\psi(\theta)$ must change by an integral multiple of 2π. If the vector at P is never directed away from the center of the disk, $\psi(\theta)$ could never equal a multiple of 2π. In this case the change of $\psi(\theta)$ over the boundary of the disk must be 0. Because $\phi(\theta) = \theta + \psi(\theta)$, the change of $\phi(\theta)$ must be 2π as θ increases from 0 to 2π. This means that the winding number of the vector field about the boundary of the disk equals 1. Because the winding number also equals the sum of the indices of the singularities, there must be at least one singularity of the vector field. This contradiction of the assumption that all vectors are nonzero proves that there must be a point P on the boundary of the disk at which the vector points away from the center.

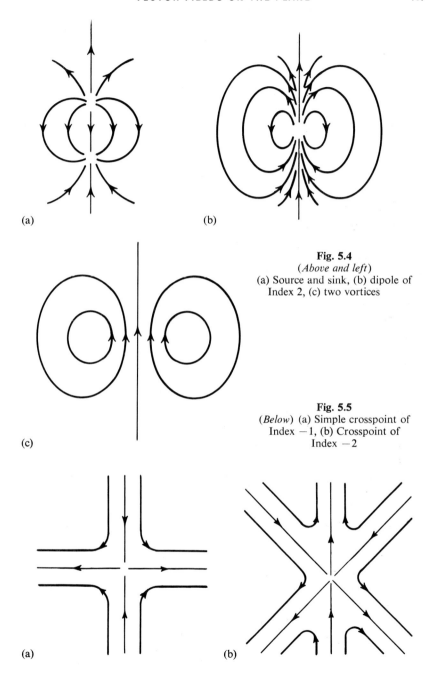

(a) (b)

(c)

Fig. 5.4
(*Above and left*)
(a) Source and sink, (b) dipole of
Index 2, (c) two vortices

Fig. 5.5
(*Below*) (a) Simple crosspoint of
Index −1, (b) Crosspoint of
Index −2

(a) (b)

5.2 A Geographical Application

Consider the plane region of points inside the counterclockwise curve e_0 and outside the clockwise curves $e_1, \ldots, e_r$ as a map of the geography of an island with r lakes. Let $e(x, y)$ be the elevation of the terrain at the point (x, y). We assume that $e(x, y)$ and its first partial derivatives are continuous. The vector $f(z)$, with $\partial e/\partial x$ and $\partial e/\partial y$ as its x- and y-coordinates, is called the gradient of e and points in the direction of steepest ascent from the point represented by $z = x + iy$ or (x, y). The gradient is a continuous vector field over the map of the island. The isolated singularities are of three types: sources at which $e(x, y)$ has a relative minimum, sinks at which $e(x, y)$ has a relative maximum, and crosspoints at which $e(x, y)$ has a saddlepoint. There can be curves of singularities such as the level rim of a crater. Geographical names for the singularities or critical points of the terrain are *peaks* for the maxima, *pits* for the minima, and *passes* or *cols* for the saddlepoints. The multiplicity of a pass is the absolute value of the index of the pass as a singularity of the gradient vector field. The second theorem in Section 5.1 has the following corollary.

Theorem. If the only critical points on the terrain of an island with r lakes are a finite number of peaks, pits, and passes in the interior of the island, the sum of the number of peaks and the number of pits diminished by the number of passes (counted with their multiplicities) equals $1 - r$.

Because the gradient at the shore of an ocean or lake must point inland, the winding number of the gradient is 1 around the counterclockwise curve e_0 and -1 around each of the clockwise curves $e_1, \ldots, e_r$. This shows that

$$\sum_{i=0}^{r} \omega(f(e_i), 0) = 1 - r.$$

The solution curves of the equations

$$\frac{dx}{dt} = \frac{\partial e}{\partial x}, \qquad \frac{dy}{dt} = \frac{\partial e}{\partial y}$$

are the curves of steepest ascent of the topography. The curve defined by $e(x, y) = c$ for a constant c is an equialtitude contour line. On this curve

$$\frac{\partial e}{\partial x} + \frac{\partial e}{\partial y}\frac{dy}{dx} = 0 \quad \text{or} \quad \frac{dy}{dx} = -\frac{\partial e/\partial x}{\partial e/\partial y}.$$

A shore line is a particular equialtitude curve. Because the slope of the curve of steepest ascent at a nonsingular point (x, y) is the negative reciprocal of the slope of the equialtitude curve at (x, y), the family of curves of steepest ascent is orthogonal to the family of equialtitude curves at all nonsingular points. The family of equialtitude curves is everywhere tangent to the vector field which assigns the vector with x- and y-coordinates $\partial e/\partial y$, $-\partial e/\partial x$ to the point (x, y). Although this vector field has the same singularities as the gradient field, the pits and peaks are now vortices rather than sources and sinks. The passes are crosspoints for both vector fields.

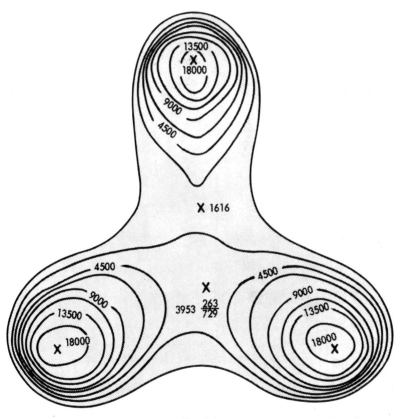

Fig. 5.6
An island on a planar map. Peaks at $(4, 0)$, $(0, 8)$, $(-4, 0)$; Passes (of Index 1) at $(0, 4/3)$ and $(0, 4)$

A relation between the number of pits, peaks, and passes was derived by Reech in a paper published in 1858. Cayley also showed the dependence of these numbers in 1859.†

Example 2. Consider an island that appears on a planar map as the set of points (x, y) such that

$$e(x, y) = 18{,}000 - [(x - 4)^2 + y^2][x^2 + (y - 8)^2][(x + 4)^2 + y^2] \geq 0.$$

Let $e(x, y)$ be the elevation of the terrain at (x, y). Figure 5.6 is a map of

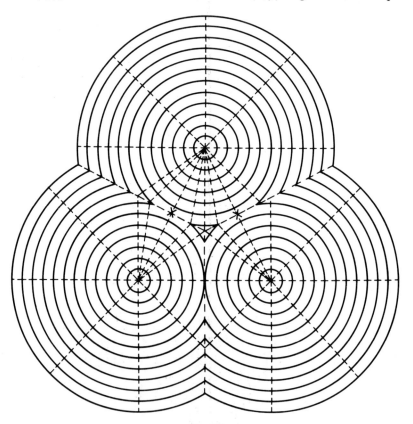

Fig. 5.7
Hypothetical island. Peaks at $(1, 0)$, $(0, 2)$, $(-1, 0)$; Passes (of Index 1) at $(0, 0)$, $(\frac{1}{2}, 1)$, $(-\frac{1}{2}, 1)$; Pit at $(0, \frac{3}{4})$

---

† Cayley's paper, with one by Maxwell, is summarized in *Mathematical Essays and Exercises* by Ball and Coxeter (New York: Macmillan, 1939) under the title *Physical Configuration of a Country.*

the island which shows equialtitude curves with interval 2250. Because there are three peaks and two passes of index 1, the formula of the theorem is satisfied.

Example 3. Consider an island that appears on a planar map as the set of points (x, y) at distance less than or equal to 2 from one or more of the points $(1, 0)$, $(0, 2)$, and $(-1, 0)$. Let the elevation of the terrain at (x, y) be

$$e(x, y) = \max(2 - \sqrt{(x-1)^2 + y^2},$$
$$2 - \sqrt{x^2 + (y-2)^2}, \, 2 - \sqrt{(x+1)^2 + y^2}).$$

Figure 5.7 is a map of the island which shows equialtitude curves (solid) and curves of steepest descent (dotted). There are peaks at $(1, 0)$, $(0, 2)$, and $(-1, 0)$, passes (of index 1) at $(\frac{1}{2}, 1)$, $(-\frac{1}{2}, 1)$, and $(0, 0)$, and a pit at $(0, \frac{3}{4})$. The equialtitude curves and the curves of steepest descent

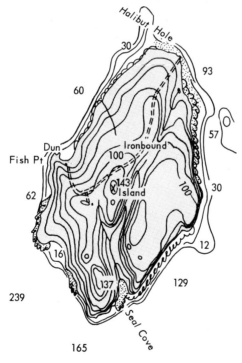

Fig. 5.8
Portion of the United States Geological Survey map of Bar Harbor, Maine

intersect at right angles except at the peaks and on the three line segments from the pit to the shore via the passes. These line segments are curves of steepest descent but not of steepest ascent. (Why?) Because the first partial derivatives of $e(x, y)$ do not exist along the three line segments, the gradient is undefined and the points of these segments are singularities of the gradient vector field. Although the lines of singularities exclude this island from the scope of the theorem, the equation relating the peaks, passes, and pits is still satisfied.

Example 4. Figure 5.8 shows a contour map of Ironbound Island, Maine. We leave to the reader the verification of the equation of the theorem.

5.3   Vector Fields and Hydrodynamics

If water flows over a plane region, the velocity vectors of the flow form a continuous vector field. The curves tangent to the vector field are the streamlines of the flow. If $u(x, y)$ and $v(x, y)$ are the coordinates of the velocity, a potential function $P(x, y)$ is a solution to the differential equations

$$\frac{\partial P}{\partial x} = -u(x, y), \qquad \frac{\partial P}{\partial y} = -v(x, y).$$

Because the gradient of $P(x, y)$ is the negative of the velocity, the streamlines are the curves of steepest descent for the potential function. The family of curves orthogonal to the streamlines is the family of equipotential lines of the flow.

From the physical assumption of the incompressibility of water, we could derive the requirement that $P(x, y)$ must be a harmonic function. A function is *harmonic* if it satisfies Laplace's differential equation

$$\frac{\partial^2 P}{\partial x^2} + \frac{\partial^2 P}{\partial y^2} = 0.$$

Every harmonic function $P(x, y)$ has a *conjugate harmonic function* $Q(x, y)$ determined by the equations

(1) $$\frac{\partial Q}{\partial x} = -\frac{\partial P}{\partial y}, \qquad \frac{\partial Q}{\partial y} = \frac{\partial P}{\partial x}.$$

When P is a potential function of a flow, P is constant along equipotential lines, whereas Q is constant along streamlines. The *conjugate flow* is defined by using Q as a potential function and the negative of the gradient of Q as the velocity. At singularities of a flow the functions P and Q need not be defined.

The complex function

$$f(z) = f(x + iy) = P(x, y) + i Q(x, y),$$

which combines a pair of conjugate harmonic functions, is called an *analytic function*. Analytic functions are complex functions, differentiable in the sense that

$$f'(z) = \lim_{\Delta z \to 0} \frac{f(z + \Delta z) - f(z)}{\Delta z}$$

exists. Comparison of the limits when $\Delta z = \Delta x$ and $\Delta z = i \Delta y$ shows that (1) must hold when $P(x, y)$ and $Q(x, y)$ are the real and imaginary parts of a differentiable complex function. These are the *Cauchy-Riemann equations* for analytic functions.

Helmholtz discovered in 1858 that the potential function of a flow over a multiply connected region might be multiple-valued, even though there were no singularities in the flow. An example is the potential function of water flowing with velocity $[-y/(x^2 + y^2), x/(x^2 + y^2)]$ around a circular annulus centered at the origin. The streamlines of this flow are concentric circles, whereas the equipotential lines are segments of rays from the origin (Figure 5.9). Because the potential function decreases as a point

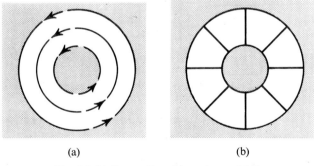

(a)   (b)

Fig. 5.9 (a) Stream lines (b) equipotential lines

moves counterclockwise along a streamline, a lower value of potential is reached each time the point returns to its initial position. For this flow

$$P(x, y) = \arg(x + iy), \qquad Q(x, y) = -\log(x^2 + y^2),$$

$$f(z) = \arg(x + iy) - i \log(x^2 + y^2) = i \log z.$$

In the conjugate flow the velocity of the water is directed toward the origin. If the inner radius of the annulus is allowed to shrink until the annulus becomes a disk, the origin is a vortex of the streamlines of the original flow and a sink of the conjugate flow. In both cases the origin is a singularity in an extended sense, for the lengths of the velocity vectors tend to infinity rather than zero as a point approaches the origin. Because the index of a singularity depends on the direction but not the length of vectors at nearby points, the definition of index can be immediately extended to the new type of singularity.

We have seen examples of a vortex and a sink. Other possible singularities of hydrodynamic flows are sources and crosspoints. In hydrodynamics the crosspoints are called *stagnation points*. We can restate the theorem of Section 5.2.

Theorem. If water flows in a lake in which there are r islands and only a finite number of singularities, with none at the shore (in particular, no streams feeding or emptying the lake), the number of sources (springs), sinks, and vortices (eddies) diminished by the number of stagnation points (counted with their multiplicities) equals $1 - r$.

5.4 Vector Fields and Differential Equations

In studying differential equations and double integrals of functions of two complex variables, Poincaré was led to topological problems. In 1895 he published "Analysis Situs," a topological paper which does not mention the motivating problems from analysis. This paper, more than a hundred pages long, together with its supplements established topology, then called *analysis situs*, as an independent branch of mathematics.

Poincaré considered the simultaneous differential equations

$$\frac{dx}{dt} = P(x, y), \qquad \frac{dy}{dt} = Q(x, y),$$

where P and Q are polynomials in x and y. Associate with the point (x, y) the vector with coordinates $P(x, y)$ and $Q(x, y)$. In the xy-plane the solution curves, called *orbits*, are everywhere tangent to the vector field. Poincaré studied the pattern of the orbits, especially the closed orbits and the singularities.

Poincaré described three types of singularity: *nodes, cols,* and *foci.* The nodes, which are sources or sinks of the vector field, and the foci have index 1, whereas the cols or crosspoints have a negative index. Because the winding number of the vector field over a positively oriented closed orbit is 1, the second theorem of Section 5.1 requires that the sum of the indices of the singularities inside the orbit equals 1, provided that the number of singularities is finite. A consequence is the following result of Poincaré.

Theorem. There is at least one singularity inside a closed orbit of the differential equations

$$\frac{dx}{dt} = P(x, y), \qquad \frac{dy}{dt} = Q(x, y),$$

where $P(x, y)$ and $Q(x, y)$, are polynomials.

We have discussed vector fields in the plane and have illustrated their applications. If a surface is embedded in space so that it has a continuously turning tangent plane, study of continuous fields of tangent vectors on the surface would lead to generalizations of the properties of vector fields on the plane; for example, Poincaré proved that the sum of the indices of a continuous tangent vector field (with a finite number of singularities) defined over a closed orientable surface equals the Euler characteristic of the surface. This implies that the torus is the only closed orientable surface that could have a continuous tangent vector field without singularities. Figure 5.10 shows that such a nonsingular vector field does exist. In his

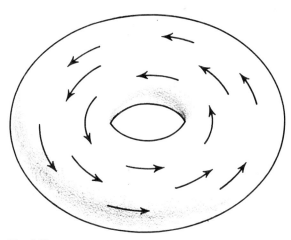

Fig. 5.10   A torus

book *On Riemann's Theory of Algebraic Functions*, Felix Klein studied hydrodynamics on a closed surface, and Marston Morse has extended Reech's results not only to surfaces but also to manifolds in n dimensions. His general theory of critical points is important in the calculus of variations.

Example 5. Let the motion of particles in the plane be determined by the differential equations

$$\frac{dx}{dt} = 9y, \qquad \frac{dy}{dt} = -4x.$$

Differentiation of the first equation gives

$$\frac{d^2x}{dt^2} = 9\frac{dy}{dt} = -36x.$$

The general solution of this equation is

$$x = c \sin 6t + d \cos 6t,$$

where c and d are arbitrary constants. Differentiation shows that

$$\frac{dx}{dt} = 6c \cos 6t - 6d \sin 6t = 9y.$$

Hence

$$y = \tfrac{2}{3}c \cos 6t - \tfrac{2}{3}d \sin 6t.$$

Now

$$(2x)^2 + (3y)^2 = (2c \sin 6t + 2d \cos 6t)^2 + (2c \cos 6t - 2d \sin 6t)^2$$
$$= 4c^2(\sin^2 6t + \cos^2 6t) + 4d^2(\cos^2 6t + \sin^2 6t)$$
$$= 4(c^2 + d^2).$$

Equivalently,

$$\frac{x^2}{c^2 + d^2} + \frac{9y^2}{4(c^2 + d^2)} = 1.$$

Thus the orbits of the system of differential equations is the family of ellipses with center at $(0, 0)$, with horizontal major axis, and with the lengths of the major and minor axes in the ratio of 3 to 2 (Figure 5.11).

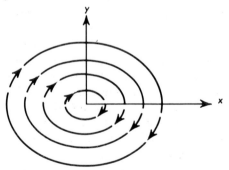

Fig. 5.11
Ellipses

The only singularity of the family of orbits is a vortex at the origin. (A vortex is a special type of focus.) Since $dy/dt < 0$, when x is positive, and $dy/dt > 0$, when x is negative, the particles travel clockwise around the origin.

5.5   Vector Fields on a Sphere

A continuous tangent vector field on a Euclidean sphere Σ is a continuous function which assigns to each point of the sphere a vector in the plane tangent to the sphere at that point. When the vector is zero, the point is a singularity. To define the index of a singularity we map the tangent vector field on the sphere onto a vector field over a plane.

If P is an isolated singularity, map the sphere by stereographic projection from a point N not equal to P onto the plane Δ tangent to Σ at N^*, the antipode of N. Introduce rectangular coordinates in Δ with N^* as the origin. Interpreting Δ as the finite part of the extended complex plane, we orient disk boundaries on Σ so that stereographic projection has Brouwer degree $+1$. Because an oriented line through a point $Q(\neq N)$ on the sphere is projected from N into an oriented line in Δ through the image of Q, stereographic projection almost defines a mapping of the tangent vector field into a planar vector field. The difficulty is that if the line segment representing the vector at Q crosses the plane Δ^*, which is tangent to Σ at N, the image of the line segment is a pair of unbounded line segments that should be connected by a projective point at ∞. These unbounded segments cannot be interpreted as a vector in Δ. The problem is solved if we use stereographic projection to determine the direction of the image of the vector at Q but let the length of the two vectors be the same. Except at N, we have defined a continuous mapping of the sphere and its tangent vector field onto the plane and a vector field on the plane. The index $I(P)$ of the singularity on the sphere is defined as the index of the image of P as a singularity of the vector field over the plane.

We wish to show that the index of P does not depend on the selection of N. Let $f(Q)$ denote the complex coordinate of the vector in Δ, which corresponds to Q. The index of P equals the winding number $\omega(f(C), 0)$, where C is the positively oriented boundary of a small disk in Σ containing P and not N or any singularity other than P. Because $f(Q)$ is a continuous function of both Q and N, $\omega(f(C), 0)$ is a continuous function of N. Now a continuous function whose values are integers must be a constant. This shows that $I(P)$ does not depend on N.

If C is the positively oriented boundary of a disk which contains only a finite number of singularities, with none on C, the winding number

$\omega(f(C), 0)$ equals the sum of the indices of the singularities in the disk. We use this statement to prove the following:

Theorem. If a continuous tangent vector field on a Euclidean sphere has only a finite number of singularities, the sum of the indices of the singularities is 2.

Let N be a point that is not a singularity. Introduce coordinates of latitude and longitude so that N is the north pole and the tangent vector at N lies over the 0-meridian. Let D be a small circular disk with N as center and C as the positively oriented circumference (Figure 5.12). Figure 5.13 shows the stereographic map of D from N^*, whereas Figure 5.14 is the stereographic map from N of the portion of the sphere from C southward. On both maps the plane vectors drawn correspond to the tangent vectors

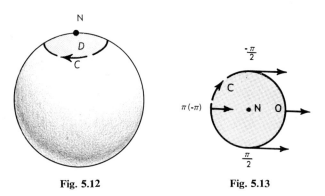

Fig. 5.12   Fig. 5.13

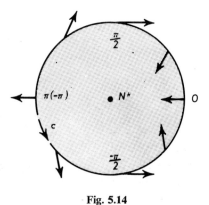

Fig. 5.14

at points of C. From the continuity of the tangent vector field we know that all the vectors in Figure 5.13 are approximately parallel if D is small enough. Now a vector that does not change direction as the image of C is traversed in Figure 5.13 corresponds to a vector that makes two revolutions as a point moves around the image of C in Figure 5.14. Thus the sum of the indices of the singularities outside D is 2. Because there are no singularities in D, the sum of the indices of the singularities on the whole sphere is 2.

Corollary. Any continuous tangent vector field on a Euclidean sphere must have a singularity.

If the members of a family of simple closed curves on a Euclidean sphere are always tangent to the vectors of a continuous tangent vector field, the corollary shows that the family cannot cover the sphere without singularities. Singularities of the family are points that are on no curve or points that are on more than one curve. This suggests the following:

Theorem. There is no family of simple closed curves on a sphere such that every point on the sphere is on one and only one curve.

A simple closed curve on a sphere could have corners or rapid oscillations that would keep the curve from having a well-defined direction. The curve defined in the plane by

$$y = x \sin \frac{1}{x}, \quad \text{if} \quad x \neq 0,$$

$$y = 0, \qquad \text{if} \quad x = 0,$$

is an example of a curve that oscillates too rapidly to have a direction at the origin (Figure 5.15). Because the directions of curves are not always defined, tangent vector fields are not weapons of sufficient power to attack this theorem.

To show the tools needed we shall outline the steps of a proof. Suppose a family of simple closed curves is on a Euclidean sphere such that each point of the sphere is on exactly one curve. By stereographic projection from the north pole onto a plane every curve of the family except the curve through the north pole is mapped onto a simple closed curve of the plane. The Jordan curve theorem states that every simple closed curve in the plane divides the plane into two regions, a bounded one inside the curve and an unbounded one outside the curve, so that no curve can join points inside to points outside without crossing the original curve. The union of

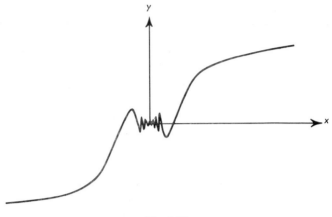

Fig. 5.15
$$y = x \sin 1/x \qquad \text{if } x \neq 0,$$
$$y = 0 \qquad \qquad \text{if } x = 0$$

the simple closed curve C and the points inside C is a closed bounded set D, which is topologically a disk with boundary curve C.

Although the Jordan curve theorem is obvious for curves like the circle or the perimeter of a regular polygon, the proof is difficult for general curves. Figure 5.16 shows a simple closed curve for which the distinction between outside and inside is less obvious than in the circle. When we consider two nonintersecting simple closed curves C_1 and C_2 in the plane, either one is entirely inside the other or each is entirely outside the other. Stated in terms of the corresponding disks, $D_1 \subset D_2$, $D_1 \supset D_2$, or the intersection $D_1 \cap D_2$ is empty. Let F be the family of simple closed curves

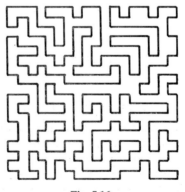

Fig. 5.16

in the plane corresponding to the curves on the sphere and let G be the corresponding family of disks. The proof of the theorem is in two parts.

Case 1. There is a disk D in G such that if D_1 and D_2 are distinct disks from G which are contained in D either $D_1 \subset D_2$ or $D_1 \supset D_2$.

In this case the intersection $D_1 \cap D_2 \cap \cdots \cap D_n$ equals D_j for some j if $D_1, D_2, \ldots, D_n$ are members of G which are contained in D. This means that the family of disks from G which are contained in D has the *finite intersection property;* namely, any intersection of a finite number of the disks is nonempty. A theorem from point set topology states that if a family of closed subsets in a bounded region of the plane has the finite intersection property the intersection of all the sets is nonempty. Let P be a point in all of the disks contained in D. If C_p is the curve from F which passes through P, select a point Q inside C_p. The curve C_Q through Q determines a disk $D_Q \subset D$ with P not in D_Q. This contradiction shows that Case 1 is impossible.

Case 2. Every disk D from G contains disks D' and D^* from G such that $D' \cap D^*$ is empty.

The definition of area given in analysis assigns a positive area to each of the disks from G. If D' and D^* are nonintersecting disks in D, at least one of them must have an area less than or equal to half the area of D. Starting with a fixed disk D from G, we can find a sequence

$$D \supset D^1 \supset D^2 \supset \cdots \supset D^n \supset \cdots$$

of disks from G such that each disk has an area less than or equal to half the area of the preceding disk. Because the sequence of disks has the finite intersection property, the Heine-Borel theorem guarantees a point P common to all disks D^n. Hence $D_P \subset D^n$ for every n. This is impossible, for area of D^n tends to zero but the area of D_P is positive.

Since neither Case 1 nor Case 2 is possible, there can be no family of simple closed curves on the sphere such that each point is on one and only one curve.

5.6   Mappings of a Sphere into Itself

The Brouwer fixed-point theorem states that every continuous mapping of a disk into itself has a fixed point. This does not generalize to mappings of a sphere into itself. An example of a continuous mapping without fixed points is the antipodal mapping of a Euclidean sphere which moves every point onto its antipode. We are able to prove that the square of any continuous mapping of the sphere into itself must have a fixed point. This and other theorems follow from a lemma.

Lemma. If f and g are continuous mappings of a sphere into itself and the product mapping gf is defined by $gf(P) = g(f(P))$, at least one of the mappings f, g, and gf has a fixed point.

Represent the sphere as a Euclidean sphere. Consider the points P, $f(P)$, and $g(f(P))$. If P and $f(P)$ coincide, P is a fixed point of f; if $f(P) = g(f(P))$, $f(P)$ is a fixed point of g; if $P = g(f(P))$, P is a fixed point of gf. If none of the mappings has a fixed point, the three distinct points $P, f(P)$, $g(f(P))$ determine a unique circle on the sphere (Figure 5.17). Let $j(P)$ be

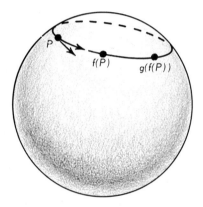

Fig. 5.17 A Euclidean sphere

the unit vector tangent to the sphere and the circular arc from P to $f(P)$ to $g(f(P))$. This defines a singularity-free, continuous, tangent vector field on the sphere. Because no such vector field can exist, there must be at least one point P for which $P, f(P)$, and $g(f(P))$ are not all distinct. At least one of the mappings f, g, and gf has a fixed point.

Theorem. If f is a continuous mapping of a sphere into itself, the mapping f^2 defined by $f^2(P) = f(f(P))$ has a fixed point.

If $g = f$ in the lemma, either f or f^2 has a fixed point. Because a fixed point of f is a fixed point of f^2, f^2 always has a fixed point.

Theorem. If f is a continuous mapping of a Euclidean sphere into itself, either f has a fixed point or f maps some point onto its antipode.

If g is the antipodal mapping, g has no fixed points, so that either f or gf has a fixed point. If gf has a fixed point P, then $g(f(P)) = f(P)^* = P$ and $P^* = f(P)$.

This theorem shows that a continuous mapping of a Euclidean sphere into itself must have a fixed point or move some point a distance equal to the diameter. If S is any topological sphere embedded in a Euclidean space, let h be a one-to-one continuous transformation of S onto a Euclidean sphere $\bar{S}$. If f is a continuous mapping of S into itself, hfh^{-1} is a continuous mapping of $\bar{S}$ into itself. If f has no fixed point, neither has hfh^{-1}. By the last theorem $(hfh^{-1})(P) = P^*$ for some point P in $\bar{S}$. Hence $f(h^{-1}(P)) = h^{-1}(P^*)$ for some point $h^{-1}(P)$ in S. Let $d(P)$ be the distance between $h^{-1}(P)$ and $h^{-1}(P^*)$. Because $d(P)$ is a continuous positive function defined on $\bar{S}$, there is a positive number m such that $d(P) \geq m$ for all P, which proves the following:

Theorem. Let S be a sphere embedded in Euclidean space. There is a positive constant m such that every continuous mapping of S into itself either has a fixed point or moves some point a distance greater than or equal to m.

This theorem shows a contrast between mappings of a circle into itself and mappings of a sphere into itself. For a circle the rotation by an angle θ with $0 < \theta < 2\pi$ about the center has no fixed points. Furthermore, for any $\varepsilon > 0$ there is an angle θ such that the points are all moved a positive distance less than ε by the corresponding rotation.

Example 5. Consider the ellipsoid defined in three-dimensional Euclidean space by the equation

$$\frac{x^2}{9} + \frac{y^2}{4} + \frac{z^2}{1} = 1.$$

The three axes of symmetry of this ellipsoid have lengths 2, 4, and 6. The function

$$h(x, y, z) = \left(\frac{x}{d}, \frac{y}{d}, \frac{z}{d}\right),$$

where $d = \sqrt{x^2 + y^2 + z^2}$, defines a one-to-one continuous mapping of the ellipsoid onto the sphere $x^2 + y^2 + z^2 = 1$. A pair of points (x, y, z) and $(-x, -y, -z)$ on the ellipsoid corresponds to a pair of antipodal points on the sphere. The shortest distance between pairs of points (x, y, z) and $(-x, -y, -z)$ on the ellipsoid is the distance 2 between $(0, 0, 1)$ and $(0, 0, -1)$. Hence any continuous mapping of the ellipsoid into itself either has a fixed point or moves some point a distance at least 2.

Map the ellipsoid onto itself by symmetry in the xy-plane so that the image of a point (x, y, z) is $(x, y, -z)$. Each point is moved a distance no greater than 2 and the points on the equator of the ellipsoid are left fixed. Follow this by a mapping that rotates the equator and nearby points slightly about the z-axis and leaves the rest of the ellipsoid fixed. The product of these mappings has no fixed point and moves no point a distance greater than 2. Therefore 2 is the largest number m such that every continuous mapping of the ellipsoid into itself either has a fixed point or moves some point a distance greater than or equal to m.

EXERCISES

Section 5.1

1. Let a continuous vector field be defined over a circle. Show that on the circle there is at least one pair of antipodal points for which the corresponding vectors have the same or opposite directions.

2. Sketch a continuous vector field over a circular disk such that the vectors on the boundary are directed away from the center and there are four singularities. Describe the singularities of your vector field.

Section 5.2

1. A survey is made of a certain island on which there are three hills and one lake (Figure 5.18). The elevations at the survey points are

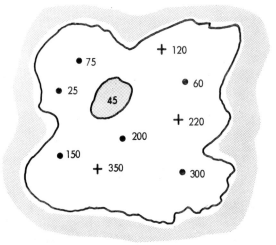

Fig. 5.18 Map of a hypothetical island

recorded on the map. Draw equialtitude contour lines at 25-foot intervals which are consistent with the given elevations. Draw on your map a path from the ocean to the top of the highest hill, to the top of the lowest hill, and back to the ocean so that the path always follows curves of steepest ascent or descent. CAUTION. Make sure that the water does not run out of the lake.

2. Sketch in the disk $x^2 + y^2 \leq 25$ equialtitude contour lines of the elevation function

$$e(x, y) = (25 - x^2 - y^2)((x + 1)^2 + (y + 1)^2)((x - 1)^2 + (y - 1)^2).$$

Describe the critical points of the terrain.

Section 5.3

1. The incomplete weather map in Figure 5.19 shows two areas of low

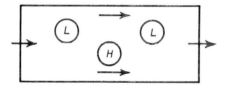

Fig. 5.19

barometric pressure and one area of high pressure. In addition, the wind is from the west at all points at the edge of the map. Complete the map by showing streamlines of the wind flow which are consistent with the given data. Label the points at which there is no wind. Sketch a second map to show the isobars (equipressure lines) that correspond to your streamlines.

2. Sketch the streamlines and equipotential lines of the flow with potential function

$$P(x, y) = \log \left| \frac{z - 1}{z + 1} \right| = \frac{1}{2} \log \left(\frac{(x - 1)^2 + y^2}{(x + 1)^2 + y^2} \right).$$

Describe the singularities at $z = 1$ and $z = -1$. HINT. An equipotential curve is the locus of points (x, y) such that the ratio of the distance of (x, y) from $(1, 0)$ to the distance of (x, y) from $(-1, 0)$ is a constant.

Section 5.4

1. If the motion of particles in the plane is determined by the differential equations

$$\frac{dx}{dt} = -2y, \qquad \frac{dy}{dt} = x - 2y,$$

show that the curve defined by the parametric equations

$$x = c_1 e^{-t} \cos t + c_2 e^{-t} \sin t,$$

$$y = \tfrac{1}{2}(c_1 - c_2)e^{-t} \cos t + \tfrac{1}{2}(c_1 + c_2)e^{-t} \sin t$$

is an orbit in which particles can travel. If a particle starts at $(1, 0)$ when $t = 0$ graph the orbit of the particle. Are there any closed orbits? Describe the singularity of the family of orbits at the origin.

2. If the motion of particles in the plane is determined by the differential equations

$$\frac{dx}{dt} = 9y, \qquad \frac{dy}{dt} = 4x,$$

find the orbits in which particles can travel. Describe the singularity of the family of orbits at the origin.

Section 5.5

1. Sketch a continuous tangent vector field on a Euclidean sphere which has three singularities.

2. Prove that there is no continuous mapping of a Euclidean sphere with center O onto itself such that each point P is mapped onto a point on the great circle whose plane is perpendicular to the radius OP. HINT. Use the image of P to define a unit vector tangent to the sphere at P.

Section 5.6

1. Prove: at least as many continuous mappings of a sphere onto itself have fixed points as have not. HINT. Use mappings with fixed points to "count" the mappings without fixed points by defining a one-to-one correspondence from the set of mappings without fixed points into the set of mappings with fixed points.

2. Prove: a continuous mapping of a Euclidean sphere into itself which maps each pair of antipodal points into a pair of distinct points maps

the sphere onto itself. HINT. If the mapping is not onto the sphere, interpret it as from a sphere to a plane. Next apply the Borsuk-Ulam theorem.

3. The image (y_1, y_2, y_2) of a point (x_1, x_2, x_3) under a linear transformation T of three-dimensional Euclidean space is determined by a set of equations:

$$y_1 = a_{11}x_1 + a_{12}x_2 + a_{13}x_3,$$

$$y_2 = a_{21}x_1 + a_{22}x_2 + a_{23}x_3,$$

$$y_3 = a_{31}x_1 + a_{32}x_2 + a_{33}x_3.$$

Show that some point (x_1, x_2, x_3) with $x_1^2 + x_2^2 + x_3^2 = 1$ is mapped on the point (cx_1, cx_2, cx_3) for some real number c. (The number c is called a characteristic root of the linear transformation.) HINT. Consider the auxiliary mapping of the sphere into sphere defined by

$$(x_1, x_2, x_3) \rightarrow \left(\frac{y_1}{d}, \frac{y_2}{d}, \frac{y_3}{d}\right)$$

where $d = \sqrt{y_1^2 + y_2^2 + y_3^2}$.

4. For the rectangular parallelepiped S with edges of length 1, 2, and 3 find the largest number m such that every continuous mapping of the surface of S into itself either has a fixed point or moves some point a distance at least as great as m.

6 NETWORK TOPOLOGY

6.1 Introduction

A relationship between persons or things may be represented by a diagram of points (vertices) and lines (edges). For example, a subway system may be sketched by representing stations by vertices and connecting successive stations on each route with an edge. The vertices of a "family tree" represent people and the edges are drawn from parents to their children. In a "game tree" each possible situation or "state" is pictured as a vertex and the edges go from one state to all other states that may be reached from it by a single move of the game. In the subway diagram no direction need be assigned to the lines, for subway trains travel in both directions. Mathematically stated the relation between successive stations is symmetric. In contrast, the edges in a family or game tree are oriented with a positive direction because the relation of parent to child is not symmetric and the play of a game is not reversible. In these examples the validity and usefulness of the diagrammatic representation of the relation does not depend on the positions of the vertices, whether the edges are straight or curved, but only on which edges meet at which vertices and what orientations are assigned to the edges. These intersections and orientations give a topological description of the network of edges and vertices.

Not all networks of edges and vertices are based on relations, for a network may have many edges connecting the same pair of vertices. A particular example of such a network was used by Euler in the Königsberg bridge problem. In Euler's time there were seven bridges at Königsberg situated as in Figure 6.1. The question was asked whether a person could

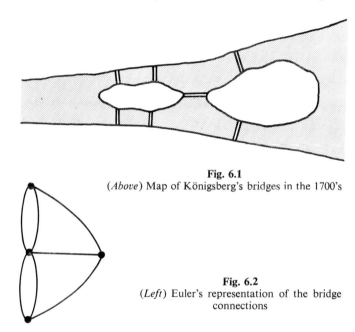

Fig. 6.1
(*Above*) Map of Königsberg's bridges in the 1700's

Fig. 6.2
(*Left*) Euler's representation of the bridge connections

walk across all seven bridges without crossing any bridge twice. Euler represented the bridge connections with a network in which the four vertices are the four land areas and the edges are the seven bridges.

The question then became whether the edges of the network could be arranged into a path that used no edge more than once. Such a path is said to be *unicursal*. We call the *degree* of a vertex the number of edges with the vertex as endpoint. Euler showed that the edges of a connected network may be arranged in a unicursal path if and only if the number of vertices with odd degree is either zero or two. If there are two vertices with odd degree, the unicursal path must start at one of them and finish at the other. If there are no vertices with odd degree, the unicursal path may start at any vertex but must finish at the same vertex. If the bridges at

Königsberg had been restricted to one way traffic, or more generally if the edges of a connected network are oriented, the question may be asked whether there is a unicursal path that uses the edges only in prescribed orientation. If the unicursal path is to finish at its starting point the number of edges oriented toward a vertex must equal the number oriented away from it. There will be exceptions to this equality at the initial and terminal points of the unicursal path if these points are distinct.

A question which was unresolved for many years is whether every map in a plane can be colored with four colors in such a way that two countries (connected polygonal plane regions) will have a different color if they have a common boundary edge. In each country select an interior point (capital). If two countries share an edge, connect their capitals by a road (edge) crossing a frontier only at a single point on the common edge. If the countries share two boundary edges, there will be two distinct roads joining the capitals. The map-coloring problem becomes one of coloring the capitals (vertices) of the road network so that no two capitals at the opposite ends of a road (edge) have the same color.

We have formulated the four-color problem in terms of planar networks. Although every network can be represented in three-space so that the curve segments representing the edges intersect only in the prescribed vertices, there are networks that cannot be so represented in the plane. This fact is the basis for the following popular problem:

Three factories are each to be connected to three utilities. No main is to pass over any other. Is it possible to satisfy these requirements?

Figure 6.3 shows how all the connections (solid lines) except the connection (dotted line) from factory A to utility III may be made. If the mains could be laid, they would divide the plane (or sphere) into polygonal faces. The number of edges would be nine, the number of vertices, six, and the Euler characteristic, 2. The number of faces would be $2 + 9 - 6 = 5$. Because the boundary of a face must cross from the factories to the utilities the same number of times as from the utilities to the factories, each of the five polygons must have an even number of edges. Because none can have only two, each must have at least four boundary edges. The number of edges is therefore at least twice the number of faces, that is, at least 10, because each edge is on the boundary of exactly two faces and at least four edges bound each face. Because there were only nine edges, we have

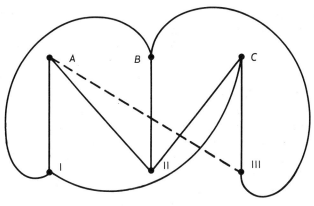

Fig. 6.3

derived a contradiction to the assumption that there is a network of mains satisfying the requirements.

Another nonplanar network is shown in Figure 6.4. Kuratowski proved that any nonplanar network contains a subnetwork identical to a subdivision of one of these two.

In this section we have sketched a few topological network problems. The following sections describe in more detail the boundary and coboundary operators in network topology and their application to two problems, one electrical, the other economic.

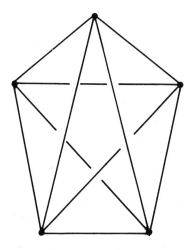

Fig. 6.4

6.2   Boundary and Coboundary

The topology of networks revolves around two operators, the boundary and the coboundary. The boundary ∂ transforms numerical functions of edges into numerical functions of vertices. The coboundary δ changes functions of vertices into functions of edges. When the boundary is used, the functions of edges or vertices are called 1-chains or 0-chains, and we represent them by column vectors. The coordinates of these column vectors are the functional values at the various edges or vertices. In the coboundary context the functions are 1-cochains and 0-cochains, represented by row vectors. If we think of each coordinate of a 1-chain as a rate of flow or current of electricity, the boundary of the 1-chain is the 0-chain whose coordinate at each vertex is the rate of accumulation or depletion of charge. When we consider a 0-cochain as a potential function, a coordinate of the 1-cochain, which is the coboundary of the 0-cochain, is the difference of potential along the corresponding edge.

When a 1-chain represents a current, a positive coordinate indicates that the direction of flow agrees with the orientation of the edge. The coordinate is negative when the flow opposes the orientation. The rate of accumulation of charge at a vertex is the algebraic sum of the currents in edges directed toward the vertex diminished by the sum of the currents in edges directed away from the vertex. For a 0-cochain representing a potential, the difference of potential over an edge is defined as the potential at the terminal point of the edge minus the potential at the initial point.

For the network in Figure 6.5 the following equations give the values of the boundary and coboundary of the 1-chain z and the 0-cochain p.

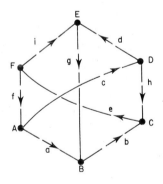

Fig. 6.5

$$\partial z(A) = -z(a) \qquad\quad - z(c) \qquad\qquad\qquad + z(f)$$
$$\partial z(B) = \quad z(a) - z(b) \qquad\qquad\qquad\qquad\quad + z(g)$$
$$\partial z(C) = \qquad\quad z(b) \qquad\qquad - z(e) \qquad\qquad + z(h)$$
$$\partial z(D) = \qquad\qquad\quad z(c) - z(d) \qquad\qquad\qquad - z(h)$$
$$\partial z(E) = \qquad\qquad\qquad\quad z(d) \qquad\qquad - z(g) \qquad + z(i)$$
$$\partial z(F) = \qquad\qquad\qquad\qquad z(e) - z(f) \qquad\qquad\qquad - z(i)$$

$$\delta p(a) = -p(A) + p(B)$$
$$\delta p(b) = \qquad\quad - p(B) + p(C)$$
$$\delta p(c) = -p(A) \qquad\qquad\quad + p(D)$$
$$\delta p(d) = \qquad\qquad\qquad - p(D) + p(E)$$
$$\delta p(e) = \qquad\quad - p(C) \qquad\qquad + p(F)$$
$$\delta p(f) = \quad p(A) \qquad\qquad\qquad\quad - p(F)$$
$$\delta p(g) = \qquad\quad p(B) \qquad\qquad + p(E)$$
$$\delta p(h) = \qquad\qquad p(C) - p(D)$$
$$\delta p(i) = \qquad\qquad\qquad\quad p(E) - p(F)$$

In the vectorial form the equations are

$$\partial z = Mz \quad \text{and} \quad \delta p = pM,$$

where M is the matrix

$$\begin{bmatrix} -1 & 0 & -1 & 0 & 0 & 1 & 0 & 0 & 0 \\ 1 & -1 & 0 & 0 & 0 & 0 & 1 & 0 & 0 \\ 0 & 1 & 0 & 0 & -1 & 0 & 0 & 1 & 0 \\ 0 & 0 & 1 & -1 & 0 & 0 & 0 & -1 & 0 \\ 0 & 0 & 0 & 1 & 0 & 0 & -1 & 0 & 1 \\ 0 & 0 & 0 & 0 & 1 & -1 & 0 & 0 & -1 \end{bmatrix}.$$

Each row of M is the vector of coefficients from the boundary equation at one vertex. Each column is the coefficient vector of the coboundary equation at one edge. The matrix entry at the intersection of a row and a column is 1, -1, or 0, according to whether the vertex corresponding to the row is the terminal point, the initial point, or not an endpoint of the edge corresponding to the column. The matrix entry corresponding to the vertex S and edge t is called the *incidence number* $\alpha(S, t)$ of the vertex and edge. The matrix M is the *incidence matrix* of the network. The individual boundary and coboundary equations have the form

$$\partial z(S) = \sum_t \alpha(S, t) z(t) \quad \text{and} \quad \delta p(t) = \sum_S p(S) \alpha(S, t).$$

The special 1-chains z for which $\partial z = Mz = 0$ are called 1-*cycles*. The 1-*coboundaries* are defined as the 1-cochains w for $w = \delta p = pM$ for some 0-chains p.

The boundary and coboundary may be used to state Maxwell's form of the Kirchhoff-Maxwell laws for a (direct current) electrical network.

I. The current 1-chain is a 1-cycle.

II. The 1-cochain whose coordinate at an edge is the impressed electromotive force in the edge diminished by the product of the current and resistance is the coboundary of the potential 0-cochain.

We refer to I as the node law and to II as the loop law. If z is the current 1-chain, w, the 1-cochain of impressed electromotive forces, R, the diagonal matrix with resistances as diagonal entries, and p, the potential 0-cochain, these laws become

$$\text{I} \quad \partial z = 0,$$

$$\text{II} \quad \delta p = w - z^T R.$$

(A vector or matrix symbol with exponent T denotes the transpose of the vector or matrix.)

To give an additional geometric interpretation of the boundary and coboundary, we identify each edge (or vertex) with the 1(0)-chain and 1(0)-cochain which has coordinate 1 corresponding to the particular edge (vertex) and has all other coordinates 0. For the network of Figure 6.5 the following are examples of equations that use this identification:

$$\partial a = -A + B, \qquad \delta A = -a - c + f.$$

The boundary equation shows that B is the terminal and A, the initial point of edge a. From the second equation it follows that a and c are the edges starting at A, whereas f is the edge finishing at A. In general, the boundary of an edge is its terminal point minus its initial point and the coboundary of a vertex is the sum of the edges terminating at the vertex diminished by the sum of the edges starting from the vertex.

Any 1-chain p with coordinates $p(t)$ may be written as the linear combination

$$p = \sum_t p(t)t.$$

Similarly, every 0-chain is a linear combination of vertices. The cochains may also be written as linear combinations.

6.3   Paths, Circuits, and Trees

We describe a path in a network by a finite sequence of edges (or inverse edges) such that the terminal point of each edge (or inverse edge) is the initial point of the next. A path is *simple* if it never passes through a vertex more than once. At most two edges in a simple path can have any particular vertex as endpoint and no two edges can be equal or inverse to each other. The *initial* and *terminal* points of a path are the initial point of the first edge and the terminal point of the last. The path is *closed* if its initial and terminal points coincide. We associate with every simple path the 1-chain which is the sum of the edges in the path, where an edge appearing inversely in the path occurs negatively in the sum. This 1-chain is called a *simple* 1-*chain*. When the simple path is not closed, the initial and terminal points of the path are initial and terminal points of the 1-chain and the boundary of the simple 1-chain is the terminal point minus the initial point. We call the simple 1-chain derived from a closed path a *circuit*. Since the boundary of a circuit vanishes, every circuit is a 1-cycle. It will be seen later that every 1-cycle is a linear combination of circuits.

A network without circuits is a tree. Figure 6.6 is a copy of Figure 6.5,

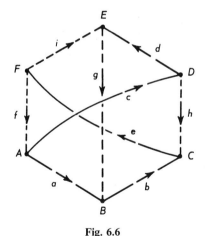

Fig. 6.6

except that some edges are dotted lines. The subnetwork of solid edges is a tree. This tree is maximal in the network, for none of the dotted edges can be added to the tree without introducing a circuit. For example, if the edge h is added, the enlarged subnetwork has the circuit $a + h - b - a$. The tree may be built up by starting with a vertex, say A, as base and some edge, say a, with A as an endpoint. The edges are added one by one, so that when an edge is added it shares exactly one endpoint with the rest of the tree. The addition of an edge cannot make a circuit, for one end of the new edge is not attached to the rest of the tree. Each time an edge is added to the tree a vertex is also added. Because the tree grew from a single edge with two endpoints, any tree will have one more vertex than it has edges. In Figure 6.6 there is a simple in-tree path (or 1-chain) from A to every other vertex. If a vertex S is left out of a tree, the tree may be enlarged by adding the first out-of-tree edge on the path from A to S. This shows that a maximal tree in Figure 6.6 must reach all six vertices. The number of edges in the maximal tree is $6 - 1 = 5$ and the number of out-of-tree edges is $9 - (6 - 1) = 4$.

The formula extended to a general network expresses the number of edges left out of a maximal tree as

$$\mu = n - m + 1,$$

where n is the total number of edges and m is the number of vertices. By *general network* we mean a collection of edges and their endpoints such that each edge has two distinct endpoints and any two endpoints (or vertices) may be joined by a simple 1-chain. This last condition requires that a network be connected. Note that because a tree is a subnetwork it is also connected. The restriction that the endpoints of an edge be distinct is to avoid special discussion of loop edges. A loop edge can always be eliminated by subdividing the loop into two edges. The number μ is the *Betti number* of the network. In Section 6.4 we shall see that all 1-cycles can be expressed in terms of μ circuits.

6.4   Basic Circuits

In the maximal tree of Figure 6.6 we select the vertex A as its *base* and say that the tree is now *rooted*. For every vertex S there is a simple, in-tree, 1-chain $\sigma(S)$ starting at A and ending at S. For example $\sigma(E) = c + d$. (For $S = A$, $\sigma(S)$ degenerates into the zero vector.) Notice that the unique simple 1-chain from E to C is

$$-\sigma(E) + \sigma(C) = -c - d + a + b.$$

We associate with any edge t the simple 1-chain $\tau(t)$ in the maximal tree which starts at the initial point of t and ends at the terminal point of t If t is not in the tree, $\omega(t) = t - \tau(t)$ is a circuit. For edges in the tree $t = \tau(t)$ so that $\omega(t) = 0$. Consider the matrices

$$
L = \begin{array}{c} \\ \\ \\ \\ \\ \\ \\ \\ \\ \end{array}
\begin{array}{cccccc}
A & B & C & D & E & F \\
\end{array}
\left[\begin{array}{cccccc}
0 & 1 & 1 & 0 & 0 & 1 \\
0 & 0 & 1 & 0 & 0 & 1 \\
0 & 0 & 0 & 1 & 1 & 0 \\
0 & 0 & 0 & 0 & 1 & 0 \\
0 & 0 & 0 & 0 & 0 & 1 \\
0 & 0 & 0 & 0 & 0 & 0 \\
0 & 0 & 0 & 0 & 0 & 0 \\
0 & 0 & 0 & 0 & 0 & 0 \\
0 & 0 & 0 & 0 & 0 & 0 \\
\end{array}\right]
\begin{array}{c}
a \\ b \\ c \\ d \\ e, \\ f \\ g \\ h \\ i \\
\end{array}
$$

$$
LM = \begin{array}{c} \\ \\ \\ \\ \\ \\ \\ \\ \\ \end{array}
\begin{array}{ccccccccc}
a & b & c & d & e & f & g & h & i \\
\end{array}
\left[\begin{array}{ccccccccc}
1 & 0 & 0 & 0 & 0 & -1 & 1 & 1 & -1 \\
0 & 1 & 0 & 0 & 0 & -1 & 0 & 1 & -1 \\
0 & 0 & 1 & 0 & 0 & 0 & -1 & -1 & 1 \\
0 & 0 & 0 & 1 & 0 & 0 & -1 & 0 & 1 \\
0 & 0 & 0 & 0 & 1 & -1 & 0 & 0 & -1 \\
0 & 0 & 0 & 0 & 0 & 0 & 0 & 0 & 0 \\
0 & 0 & 0 & 0 & 0 & 0 & 0 & 0 & 0 \\
0 & 0 & 0 & 0 & 0 & 0 & 0 & 0 & 0 \\
0 & 0 & 0 & 0 & 0 & 0 & 0 & 0 & 0 \\
\end{array}\right]
\begin{array}{c}
a \\ b \\ c \\ d \\ e. \\ f \\ g \\ h \\ i \\
\end{array}
$$

The columns of L and LM are the simple 1-chains $\sigma(S)$ and $\tau(t)$. The last rows of L and LM are zero vectors because these simple 1-chains are in the tree. The top left corner of LM is an identity matrix, for $\tau(t) = t$ for in-tree edges. The reason that the columns of LM are the vectors $\tau(t)$ is that

$$\tau(t) = \sum_{S} \sigma(S)\, \alpha(S, t).$$

If N is the matrix with columns $\omega(t)$,

$$
N = \begin{array}{c} \\ \\ \\ \\ \\ \\ \\ \\ \\ \end{array}
\begin{array}{ccccccccc}
a & b & c & d & e & f & g & h & i \\
\end{array}
$$

$$
N = \left[\begin{array}{ccccccccc}
0 & 0 & 0 & 0 & 0 & 1 & -1 & -1 & 1 \\
0 & 0 & 0 & 0 & 0 & 1 & 0 & -1 & 1 \\
0 & 0 & 0 & 0 & 0 & 0 & 1 & 1 & -1 \\
0 & 0 & 0 & 0 & 0 & 0 & 1 & 0 & -1 \\
0 & 0 & 0 & 0 & 0 & 1 & 0 & 0 & 1 \\
0 & 0 & 0 & 0 & 0 & 1 & 0 & 0 & 0 \\
0 & 0 & 0 & 0 & 0 & 0 & 1 & 0 & 0 \\
0 & 0 & 0 & 0 & 0 & 0 & 0 & 1 & 0 \\
0 & 0 & 0 & 0 & 0 & 0 & 0 & 0 & 1 \\
\end{array}\right]
\begin{array}{l}
a \\ b \\ c \\ d \\ e. \\ f \\ g \\ h \\ i
\end{array}
$$

The matrix may be obtained by subtracting LM from the 9×9 identity matrix. The first five columns of N are zero vectors because $\omega(t) = 0$ for in-tree edges. The last four columns are the circuits $\omega(t)$ for the four out-of-tree edges. The identity matrix in the lower right corner of N reflects the fact that each out-of-tree edge t is in only its own circuit $\omega(t)$.

If z is a 1-cycle, that is if $Mz = 0$,

$$Nz = z - (LM)z = z.$$

The equation

$$z = Nz \tag{1}$$

expresses the 1-cycle as a linear combination of the μ circuits $\omega(t)$ corresponding to out-of-tree edges. Because the left-hand side of equation (1) depends only on the out-of-tree coordinates of z, equation (1) gives the solution for all coordinates of z in terms of μ out-of-tree coordinates. Another form of equation (1) is

$$z = \sum_t z_t \, \omega(t),$$

which shows that any 1-cycle is a linear combination of the μ nonzero circuits $\omega(t)$ corresponding to the out-of-tree edges.

The matrix relations just developed may be used to derive a new description for the 1-coboundaries of a network. If w is a 1-coboundary,

$$w = \delta p = pM$$

for some 0-cochain p. Therefore $wz = (\delta p)z = (pM)z = p(Mz) = p(\partial z)$. For all 1-cycles z, $wz = 0$. Because the columns of N are 1-cycles,

$$wN = 0. \tag{2}$$

If, on the other hand, w is a 1-cochain such that $wN = 0$,

$$w = w(LM) = (wL)M = \delta(wL).$$

This shows that a 1-cochain w is a 1-coboundary if and only if $wN = 0$.

We shall partition the 1-chain z and the 1-cochain w into the forms

$$z = \begin{bmatrix} x \\ y \end{bmatrix} \quad \text{and} \quad w = [u, v],$$

where x and u are the subvectors with coordinates for each in-tree edge, whereas y and v have coordinates for the out-of-tree edges. The matrix LM is also written in partitioned form as

$$LM = \begin{bmatrix} I & K \\ 0 & 0 \end{bmatrix},$$

where I is a 5×5 identity matrix, K is a 5×4 matrix, and the zeros represent zero matrices with the appropriate dimensions to fill out LM. The analogous form for N is

$$N = \begin{bmatrix} 0 & -K \\ 0 & J \end{bmatrix},$$

where J is a 4×4 identity matrix. Equation 1 may be rewritten as

$$\begin{bmatrix} x \\ y \end{bmatrix} = \begin{bmatrix} 0 & -K \\ 0 & J \end{bmatrix} \begin{bmatrix} x \\ y \end{bmatrix}$$

or

$$\begin{bmatrix} x \\ y \end{bmatrix} = \begin{bmatrix} -Ky \\ y \end{bmatrix}.$$

This is equivalent to

$$x + Ky = 0. \tag{3}$$

Equation 2 may be replaced by

$$[u, v] \begin{bmatrix} 0 & -K \\ 0 & J \end{bmatrix} = 0$$

or

$$-uK + v = 0. \tag{4}$$

The solutions of the dual equations (3) and (4) are the 1-cycles and 1-coboundaries, respectively.

Although we have discussed a specific network, the reasoning used applies to general networks.

6.5   The Kirchhoff-Maxwell Laws

The Kirchhoff form of the loop law states that the sum over any circuit (loop) of the impressed electromotive forces equals the corresponding sum of the products of current times resistance. The sum of a 1-cochain, which is a row vector, over a circuit, which is a column vector, may be computed by multiplying the row vector times the column vector. Since all circuits can be expressed in terms of the basic circuits which are columns of N, the Kirchhoff form of the law is equivalent to the equation

$$\text{II*} \qquad wN = z^T RN$$

where w is the 1-cochain of impressed electromotive forces, R is the resistance matrix, and z is the current 1-cycle.

No potential 0-chain p appears explicitly in the Kirchhoff form of the loop law. Any 0-cochain p such that

$$\delta p = w - z^T R$$

is a suitable potential 0-chain. When II* is satisfied, there is always such a p, for

$$(w - z^T R)N = 0$$

means that $w - z^T R$ is a 1-coboundary. A particular potential 0-cochain is

$$p = (w - z^T R)L.$$

Other potential 0-cochains are obtained by adding a constant to all entries of p. We see that there are no other potential 0-cochains because all the potential differences are specified by the loop law.

An important electrical problem is the determination of the current 1-chain z and a potential 0-cochain p when the 1-cochain w of impressed electromotive forces and the resistance matrix R are given. The Kirchhoff-Maxwell laws supply the equations

$$z = Nz \qquad \text{and} \qquad z^T RN = wN.$$

In terms of x, y, u, and v these equations become

$$x + Ky = 0,$$

$$-x^T R_1 K + y^T R_2 = -uK + v,$$

where R_1 and R_2 are the diagonal resistance matrices for the in-tree and out-of-tree edges. These two equations give $n = \mu + (n - \mu)$ linear equations for the n unknown coordinates of x and y. To solve these equations we first take the transpose of the second matrix equation

$$-K^T R_1 x + R_2 y = (-uK + v)^T.$$

Substitution of the value of x from the first equation gives

$$(K^T R_1 K + R_2)y = (-uK + v)^T.$$

The square $\mu \times \mu$ matrix $P = K^T R_1 K + R_2$ can be shown to have an inverse when the resistances are all positive.† The solution for y is

$$y = P^{-1}(-uK + v)^T.$$

Of course, this solution may be derived in numerical cases by use of solution methods that do not require the explicit calculation of an inverse matrix.

We may now derive a formula for z in terms of w by using the equations

$$z = \begin{bmatrix} x \\ y \end{bmatrix} = \begin{bmatrix} -Ky \\ y \end{bmatrix} = \begin{bmatrix} -K \\ J \end{bmatrix} y,$$

$$-uK + v = [u, v]\begin{bmatrix} -K \\ J \end{bmatrix} = w\begin{bmatrix} -K \\ J \end{bmatrix}.$$

This formula is

$$z = \begin{bmatrix} -K \\ J \end{bmatrix} P^{-1}[-K^T, J]w^T$$

or

$$z = \begin{bmatrix} KP^{-1}K^T & -KP^{-1} \\ -P^{-1}K^T & P^{-1} \end{bmatrix} w^T.$$

Substitution in the equation

$$p = (w - z^T R)L$$

---

† Since the quadratic form $x^T R_1 x$ is positive definite,

$$(Ky)^T R_1 (Ky) = y^T (K^T R_1 K)y \geq 0$$

for all y. Therefore

$$y^T (K^T R_1 K + R_2)y = y^T (K^T R_1 K)y + y^T R_2 y > 0 \qquad \text{for all } y \neq 0.$$

Because $K^T R_1 K + R_2$ is positive definite, it is nonsingular and has an inverse.

produces the following formula for a potential 0-cochain

$$p = w \begin{bmatrix} I - KP^{-1}K^TR_1 & KP^{-1}R_2 \\ P^{-1}K^TR_1 & J - P^{-1}R_2 \end{bmatrix} L.$$

Because the out-of-tree rows of L are zero-vectors, the right-hand blocks of the partitioned matrix are multiplied by a zero matrix in L. Therefore the formula p can be simplified to

$$p = w \begin{bmatrix} I - KP^{-1}K^TR_1 & 0 \\ P^{-1}K^TR_1 & 0 \end{bmatrix} L.$$

To give an illustrative numerical example, using the network of Figure 6.6, we assign resistance 1 to edges a, c, e, g, i and resistance 2 to edges b, d, f, h. Then

$$P = \begin{bmatrix} 6 & -1 & -3 & 4 \\ -1 & 5 & 2 & -4 \\ -3 & 2 & 6 & -4 \\ 4 & -4 & -4 & 8 \end{bmatrix},$$

$$P^{-1} = \frac{1}{328} \begin{bmatrix} 96 & -32 & 24 & -52 \\ -32 & 120 & -8 & 72 \\ 24 & -8 & 88 & 28 \\ -52 & 72 & 28 & 117 \end{bmatrix},$$

$$z = \frac{1}{328} \left[\begin{array}{ccccc|cccc} 117 & 45 & -65 & -37 & 17 & 52 & -72 & -28 & -35 \\ 45 & 93 & -25 & 11 & 57 & 20 & 48 & -36 & 37 \\ -65 & -25 & 109 & 57 & 27 & 44 & 40 & 52 & -17 \\ -37 & 11 & 57 & 93 & -25 & 20 & 48 & -36 & -45 \\ 17 & 57 & 27 & -25 & 109 & 44 & 40 & 52 & 65 \\ \hline 52 & 20 & 44 & 20 & 44 & 96 & -32 & 24 & -52 \\ -72 & 48 & 40 & 48 & 40 & -32 & 120 & -8 & 70 \\ -28 & -36 & 52 & -36 & 52 & 24 & -8 & 88 & 28 \\ -35 & 37 & -17 & -45 & 65 & -52 & 72 & 28 & 117 \end{array} \right] w^T,$$

$$p = \frac{1}{328} w \begin{bmatrix} 0 & 211 & 121 & 65 & 139 & 104 \\ 0 & -45 & 97 & 25 & 3 & 40 \\ 0 & 65 & 115 & 219 & 105 & 88 \\ 0 & 37 & 15 & -57 & 85 & 40 \\ 0 & -17 & -131 & -27 & 23 & 88 \\ 0 & -52 & -92 & -44 & -84 & -136 \\ 0 & 72 & -24 & -40 & -136 & -64 \\ 0 & 28 & 100 & -52 & 20 & 48 \\ 0 & 35 & -39 & -17 & 107 & -104 \end{bmatrix}$$

The first column of the 9×9 matrix is a zero vector because p is the particular potential 0-chain with zero potential at the vertex A, which is the base of the tree.

6.6   A Transportation Problem

In the network in Figure 6.7 the vertices A and B represent warehouses from which a commodity is to be shipped to the markets represented by the vertices C, D, and E. Each edge is a shipping route from a warehouse to a market. The positive direction of the edge is from the warehouse to the market. The 0-chain

$$q = \begin{bmatrix} -55 \\ -45 \\ 50 \\ 35 \\ 15 \end{bmatrix} \begin{matrix} A \\ B \\ C \\ D \\ E \end{matrix}$$

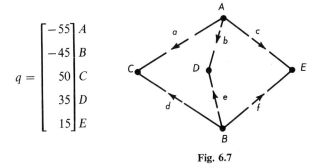

Fig. 6.7

gives shipping requirements. The positive coordinate at a market is the number of units needed at that market, whereas the negative coordinate for a warehouse shows the number of units to be shipped from that warehouse. Because the sum of the coordinates of q is zero, the supplies at the warehouses equal the demands at the markets. The costs per unit shipped along the various routes are the coordinates of the 1-cochain

$$\begin{array}{cccccc} a & b & c & d & e & f \end{array}$$
$$w_0 = [4 \quad 8 \quad 5 \quad 3 \quad 6 \quad 2].$$

We shall solve the problem of determining how the commodity should be shipped from the warehouses to the markets if the total shipping cost is to be minimized. In studying this problem we shall discover a dual problem that can be solved simultaneously.

We define a *distribution vector* for the transportation problem as a 1-chain such that the requirements at the vertices are satisfied when the number of units shipped along an edge is the corresponding chain co-ordinate. We say a distribution vector is *feasible* if none of its coordinates is negative. An *optimal distribution vector* is a feasible distribution vector that minimizes the total shipping cost. Our problem is to find an optimal distribution vector.

The incidence matrix of the network of Figure 6.7 is

$$
M = \begin{array}{c} \\ \\ \\ \\ \\ \end{array}
\begin{array}{cccccc}
a & b & c & d & e & f \\
\end{array}
\left[
\begin{array}{cccccc}
-1 & -1 & -1 & 0 & 0 & 0 \\
0 & 0 & 0 & -1 & -1 & -1 \\
1 & 0 & 0 & 1 & 0 & 0 \\
0 & 1 & 0 & 0 & 1 & 0 \\
0 & 0 & 1 & 0 & 0 & 1 \\
\end{array}
\right]
\begin{array}{c}
A \\ B \\ C. \\ D \\ E
\end{array}
$$

We select the rooted maximal tree with base A and edges a, b, c, d. The matrix L, whose columns are the simple in-tree chains from A to the various vertices, is

$$
L = \begin{array}{c} \\ \\ \\ \\ \\ \\ \end{array}
\begin{array}{ccccc}
A & B & C & D & E \\
\end{array}
\left[
\begin{array}{ccccc}
0 & 1 & 1 & 0 & 0 \\
0 & 0 & 0 & 1 & 0 \\
0 & 0 & 0 & 0 & 1 \\
0 & -1 & 0 & 0 & 0 \\
0 & 0 & 0 & 0 & 0 \\
0 & 0 & 0 & 0 & 0 \\
\end{array}
\right]
\begin{array}{c}
a \\ b \\ c. \\ d \\ e \\ f
\end{array}
$$

The product of M times a column of L is the boundary of the simple chain from A to a particular vertex. This explains the special form of

$$ML = \begin{bmatrix} 0 & -1 & -1 & -1 & -1 \\ 0 & 1 & 0 & 0 & 0 \\ 0 & 0 & 1 & 0 & 0 \\ 0 & 0 & 0 & 1 & 0 \\ 0 & 0 & 0 & 0 & 1 \end{bmatrix}.$$

A distribution vector of the transportation problem is a 1-chain z that satisfies the equation

$$Mz = q.$$

Because the first coordinate of q is the negative of the sum of the other coordinates,

$$MLq = q.$$

This shows that

$$z_0 = Lq = \begin{bmatrix} 5 \\ 35 \\ 15 \\ 45 \\ 0 \\ 0 \end{bmatrix}$$

is a particular distribution vector. This distribution is derived when shipments are restricted to the maximal tree. If z is any distribution vector and $z^* = z - z_0$,

$$Mz^* = Mz - Mz_0 = 0.$$

On the other hand, $z = z^* + z_0$ is a distribution vector whenever z^* is a 1-cycle.

A "dual" pricing problem intimately related to the transportation problem is to find among the 1-cochains that satisfy the conditions $w = w^* + w_0$ for some 1-coboundary w^* and w has non-negative coordinates the particular 1-cochain that minimizes wz_0. If p is a 0-cochain with $\delta p = -w^*$, we think of a coordinate of p as a "shadow price" or "locational value" at a vertex for a unit of the commodity being transported. After the commodity has been shipped from a warehouse to a market, the price or value increases by an amount that reflects the shipping costs. The increase along an edge is less than or equal to the shipping cost along that edge. The 1-cochain, $w = w^* + w_0$, gives the shipping cost per unit for each edge which is not balanced by an increase in the shadow price. The problem of minimizing wz_0 is simply that of assigning locational values, hence of determining w^* and w, so that the total of unbalanced shipping costs for the particular distribution vector z_0 is a minimum. We define a *loss vector* as a vector w of the form $w = w^* + w_0$, where w^* is a 1-coboundary. If all coordinates of w are non-negative, the loss vector will be feasible. We call a feasible loss vector that gives the minimum value of wz_0 an *optimal loss vector*. Although the physical problem requires z_0 to be feasible, we now find that mathematically the pricing problem has the same optimal loss vectors, no matter what distribution, feasible or not, is used for z_0.

For any distribution vector $z = z^* + z_0$,

$$wz = w(z^* + z_0) = (w^* + w_0)z^* + wz_0 = w_0z^* + wz_0,$$

because the product w^*z^* of a 1-cycle and 1-coboundary is zero. This formula shows that when z is fixed wz and wz_0 as functions over the feasible loss vectors attain their minima at the same values of w; hence the optimal loss vectors do not depend on the selection of z_0. The cost vector w_0 is the feasible loss vector obtained by making all shadow prices equal. For any loss vector $w = w^* + w_0$,

$$wz = (w^* + w_0)z = w^*(z^* + z_0) + w_0z = w^*z_0 + w_0z.$$

This shows that wz and w_0z, as functions over the feasible distribution vectors, attain their minima for the same value of z. Thus the optimal distribution vectors do not depend on which loss vector is the cost vector. If we interpret a negative coordinate as a subsidy for using a particular shipping route, we may even think of a nonfeasible loss vector as the cost vector for a transportation problem with the same optimal distribution vectors.

When w and z are feasible loss and distribution vectors,

$$wz \geq 0.$$

If w_1 and z_1 are feasible loss and distribution vectors such that

$$w_1z_1 = 0,$$

w_1 is an optimal loss vector and z_1 is an optimal distribution vector. These vectors are optimal because w_1 gives the minimum of wz_1 and z_1 the minimum of w_1z. For each maximal tree there is a distribution vector z and a loss vector w with the out-of-tree coordinates zero in z and the in-tree coordinates zero in w. For these vectors $wz = 0$. If we find a maximal tree such that the corresponding vectors w and z are both feasible, these vectors are optimal. We limit our discussion of the existence of such feasible vectors to an illustrative example.

Consider transportation problems in which goods are to be distributed from m_1 warehouses to m_2 markets via $n = m_1m_2$ routes, one directed from each warehouse to each market. These problems have optimal distribution vectors which have zero coordinates for the edges left out of some maximal tree. If z_1 is an optimal distribution vector with out-of-tree coordinates zero, let w_1 be the loss vector with zero coordinates for the edges of the same tree. The vector w_1 is feasible, hence optimal. The pricing problems over the same type of network always have an optimal loss vector with zero coordinates for the edges of a maximal tree. If w_1 is such an optimal loss vector, the corresponding distribution vector z_1 with out-of-tree coordinates zero is both feasible and optimal. Discussion of our specific transportation problem and its dual pricing problem illustrates the principle but not the details of the general proof of these facts.

For the problem on the network of Figure 6.7 we found the distribution vector

$$
z_0 = \begin{bmatrix} 5 \\ 35 \\ 15 \\ 45 \\ 0 \\ 0 \end{bmatrix}
$$

with coordinates zero for the edges e and f, which are left out of the maximal tree with edges a, b, c, and d. We write the general distribution vector as

$$
z = \begin{bmatrix} x \\ y \end{bmatrix}
$$

where x and y are the in-tree and out-of-tree parts. Because $z - z_0$ is a cycle,

$$\begin{bmatrix} x_a \\ x_b \\ x_c \\ x_d \end{bmatrix} - \begin{bmatrix} 5 \\ 35 \\ 15 \\ 45 \end{bmatrix} + \begin{bmatrix} -1 & -1 \\ 1 & 0 \\ 0 & 1 \\ 1 & 1 \end{bmatrix} \begin{bmatrix} y_e \\ y_f \end{bmatrix} = 0.$$

This matrix equation represents the four equations

$$
\begin{aligned}
x_a &= y_e + y_f + 5, \\
x_b &= -y_e + 35, \\
x_c &= - y_f + 15, \\
x_d &= -y_e - y_f + 45.
\end{aligned}
$$

Because a distribution vector is completely determined by the values of y_e and y_f, we represent the distribution vectors by the points in a plane with Cartesian coordinates (y_e, y_f). The feasible distribution vectors correspond to the points in the plane that satisfy the inequalities

$$
\begin{aligned}
y_e + y_f + 5 &\ge 0, \\
-y_e + 35 &\ge 0, \qquad y_e \ge 0, \\
 - y_f + 15 &\ge 0, \qquad y_f \ge 0, \\
-y_e - y_f + 45 &\ge 0.
\end{aligned}
$$

Figure 6.8 shows the feasible region of points that satisfies these inequalities.

The function to be minimized over the feasibility region is

$$w_0 z = 4x_a + 8x_b + 5x_c + 3x_d + 6y_e + 2y_f = -y_e - 2y_f + 510.$$

The graph of this function in three dimensions is the portion of a plane over the feasible region. The function must attain its minimum at one of the vertices of the feasible region. If the plane of the graph were parallel to one of the boundary lines of the feasible region, the function $w_0 z$ might assume its minimum on an entire boundary edge. The vertices of the feasible region are the intersections of two lines, on each of which one coordinate of z is zero. The points on one of these lines represent the distribution vectors, with none of the commodity sent along a particular edge. There are 15 pairs of lines. The three nonintersecting pairs correspond to pairs of edges which are not the out-of-tree edges for any maximal tree. Each of the 12 intersecting pairs of lines corresponds to a pair of edges

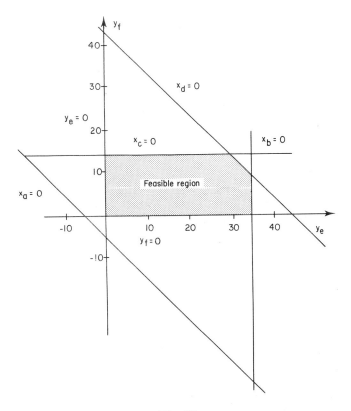

Fig. 6.8

which are the out-of-tree edges of a maximal tree. The point of intersection represents the distribution vector, with zero coordinates for these out-of-tree edges. The five points of intersection which represent feasible distribution vectors are the vertices of the feasible region. The minimum cost is attained for one of the five feasible distribution vectors.

For each maximal tree there is a 1-coboundary w^* which has the same in-tree coordinates as $-w_0$. The loss vector $w = w^* + w_0$ has zero in-tree coordinates. For each maximal tree Table 6.1 lists the distribution vector

Table 6.1

In-tree Edges	z	w^T	p^T	$w_0 z$	$w z_0$
a	5	0	0	510	
b	35	0	1		
c	15	0	4		
d	45	0	8		
	0	−1	5		
	0	−2			
a	50	0	0		90
b	35	0	1		
c	−30	0	4		
	0	2	8		
	0	1	5		
f	45	0			
a	50	0	0	450	60
b	5	0	2		
	0	1	4		
	0	1	8		
e	30	0	4		
f	15	0			
a	50	0	0	455	
	0	−1	3		
c	5	0	4		
	0	2	9		
e	35	0	5		
f	10	0			
	0	−1	0		
b	40	0	2		
c	15	0	5		
d	50	0	8		
e	−5	0	5		
	0	−1			
	0	−1	0		
b	55	0	2		
	0	1	5		
d	50	0	8		
e	−20	0	4		
f	15	0			

Table 6.1

In-tree Edges	z	w^T	p^T	$w_0 z$	$w z_0$
a	50	0	0		
b	−10	0	2		
c	15	0	4		
	0	1	8		
e	45	0	5		
	0	−1			
a	20	0	0	480	
b	35	0	1		
	0	2	4		
d	30	0	8		
	0	−1	3		
f	15	0			
a	40	0	0	475	
	0	1	1		
c	15	0	4		
d	10	0	7		
e	35	0	5		
	0	−2			
a	55	0	0		65
	0	1	1		
	0	2	4		
d	−5	0	7		
e	35	0	3		
f	15	0			
	0	−2	0		
b	35	0	3		
c	20	0	6		
d	50	0	8		
	0	1	5		
f	−5	0			
	0	−2	0		
	0	−1	3		
c	55	0	6		
d	50	0	9		
e	35	0	5		
f	−40	0			

z with zero out-of-tree coordinates, the loss vector $w = w^* + w_0$ with zero in-tree coordinates, and a vector p of shadow prices or locational values such that $\delta p = -w^*$.

The columns of values of $w_0 z$ and $w z_0$ are filled only when z or w is feasible. We notice that the tree with edges a, b, c, f is the only tree for which both z and w are feasible. The values z_1 and w_1 of z and w corresponding to this tree give the minima of $w_0 z$ and $w z_0$. The vector [0 2 4 8 4] is a vector of shadow prices such that the price increase along each edge equals the shipping cost, except on the out-of-tree edges, c and d, along which none of the commodity is shipped under the optimal distribution vector. These shadow prices would enforce the optimal distribution vector by penalizing the shipper for using route c or d.

To explore the relation between the dual problems, we shall forget that we know w_1 explicitly and prove that w_1 is feasible and optimal because z_1 is optimal. Now w_1 has the form [0 0 $w_1(c)$ $w_1(d)$ 0 0]. Because z_1 is optimal, the value $w_1 z_1 = 0$ is the minimum of $w_1 z$ as a function over the feasible region. We derive a new distribution vector z_2 from z_1 by changing the coordinate of c from 0 to 1 and making compensating changes in the positive coordinates of z_1. The change from z_1 to z_2 may be achieved by adding to z_1 the circuit

$$
\begin{bmatrix} 0 \\ -1 \\ 1 \\ 0 \\ 1 \\ -1 \end{bmatrix} \text{ to get } z_2 = \begin{bmatrix} 50 \\ 4 \\ 1 \\ 0 \\ 31 \\ 14 \end{bmatrix}
$$

Now z_2 is feasible and $z_2 w_1 = w_1(c)$. Therefore $w_1(c) \geq 0$.

$$
\text{If the circuit } \begin{bmatrix} -1 \\ 1 \\ 0 \\ 1 \\ -1 \\ 0 \end{bmatrix} \text{ is added to } z_1 \text{ to give } z_3 = \begin{bmatrix} 49 \\ 6 \\ 0 \\ 1 \\ 29 \\ 15 \end{bmatrix}, z_3 \text{ is feasible}
$$

and $w_1 z_3 = w_1(d) \geqslant 0$. This shows that w_1 is feasible. We proved earlier that if distribution vectors and loss vectors z_1, w_1 were both feasible and

$w_1 z_1 = 0$, both vectors are optimal. Dual reasoning shows that the optimality of w_1 implies the feasibility and optimality of z_1.

In conclusion, we shall relate the dual transportation and pricing problems to a pair of dual linear programming problems. We have seen that the optimum vectors do not change if we alter z_0 by a 1-cycle or w_0 by a 1-coboundary. Therefore we may assume that z_0 has zero out-of-tree coordinates and w_0 has zero in-tree coordinates with respect to a particular maximal tree. We partition the distribution vectors z and z_0, the loss vectors w and w_0, and the 1-cycles and 1-coboundaries z^* and w^* into in-tree and out-of-tree parts.

$$z = \begin{bmatrix} x \\ y \end{bmatrix}, \ z_0 = \begin{bmatrix} x_0 \\ 0 \end{bmatrix}, \ z^* = \begin{bmatrix} x^* \\ y^* \end{bmatrix}, \ w = [u \quad v], \ w_0 = [0 \quad v_0], \ w^* = [u^* v^*].$$

Formulated in terms of the 1-cycles z^* and the 1-coboundaries w^*, the dual problems are combined into the following:

Transportation Problem. Minimize $[0 \quad v_0] \begin{bmatrix} x^* \\ y^* \end{bmatrix}$ subject to the conditions

$$\begin{bmatrix} x^* \\ y^* \end{bmatrix} \geq - \begin{bmatrix} x_0 \\ 0 \end{bmatrix} \quad \text{and} \quad x^* + Ky^* = 0.$$

We use the vector inequality

$$\begin{bmatrix} a_1 \\ \vdots \\ a_n \end{bmatrix} \geq \begin{bmatrix} b_1 \\ \vdots \\ b_n \end{bmatrix}$$

as an abbreviated form of the system of simultaneous inequalities

$$a_1 \geq b_1, \ldots, a_n \geq b_n.$$

Pricing Problem. Minimize $[u^* \quad v^*] \begin{bmatrix} x_0 \\ 0 \end{bmatrix}$ subject to the conditions $[u^* \quad v^*] \geq -[0 \quad v_0]$ and $-u^*K + v^* = 0$.

We restate these problems after eliminating the dependent vectors x^* and v^* and setting $v_0 = -\bar{v}_0$.

Transportation Problem. Maximize $\bar{v}_0 y^*$ subject to the conditions $y^* \geq 0$ and $Ky^* \leq x_0$.

Pricing Problem. Minimize $u^* x_0$ subject to the conditions $u^* \geq 0$ and $u^* K \geq \bar{v}_0$.

This is a standard form for presenting a pair of dual linear programming problems. Linear programming problems may be solved by a very efficient computational algorithm (devised in 1947 by G. B. Dantzig), called the "simplex method." The algorithm is related to the above tree-analysis of the transportation problem and its dual.

EXERCISES

Section 6.1

1. A police captain ordered a policeman to patrol the 20 miles of streets drawn on the map in Figure 6.9. The captain observed the mileage indicator of the police car both before the policeman left and after he returned. When the captain saw that the police car had traveled exactly 20 miles, he suspended the policeman. Justify the suspension.

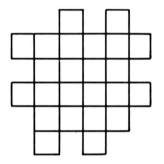

Fig. 6.9
Street diagram

2. Solve the same problem for the one-way streets on the map in Figure 6.10.

3. A chessboard has 64 squares arranged into eight rows and eight columns. A knight may move from the square in the ith row and the jth column to the square in the kth row and mth column if and only if

$$0 < |i - k| < |i - k| + |j - m| = 3.$$

Consider the 64 squares as the vertices of a network in which pairs of vertices are connected by an edge if and only if a knight may move from one square of the pair to the other. Can a knight make a unicursal trip over this network?

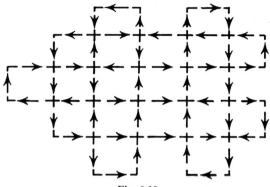

Fig. 6.10
Diagram of one-way streets

4. One or more airplanes are to be chartered to deliver 24 cargoes between four cities, *A*, *B*, *C*, and *D*. The origins and destinations of the cargoes are given in tabular form. One flight may be made per plane per day.

From	To	Number of Cargoes	From	To	Number of Cargoes	From	To	Number of Cargoes
A	*B*	3	*A*	*D*	2	*B*	*D*	2
B	*A*	2	*D*	*A*	1	*D*	*B*	2
A	*C*	2	*B*	*C*	2	*C*	*D*	1
C	*A*	4	*C*	*B*	1	*D*	*C*	2

Planes may be chartered in any city but each must be returned to the city in which it originated. In each of the following cases prepare a schedule of flights that minimizes the cost.

a. Planes rent at $25,000 per month with an additional charge of $2000 per flight.
b. Planes rent at $10,000 per flight. A penalty of $500 per cargo must be paid each day for all undelivered cargoes.

Section 6.2

1. Show that the sum of the values of a 0-chain at the various vertices equals zero if the 0-chain is the boundary of a 1-chain.

2. Show that the rank of the incidence matrix of a network is less than the number of vertices.

Section 6.3

1. Show that the coboundary of a 0-cochain on a connected network is the 1-cochain with value zero at each edge if and only if the 0-chain has the same value at every vertex.

2. Find the Betti number of the network of edges and vertices of a regular icosahedron.

Section 6.4

1. Give a set of four linear equations which are satisfied by the coordinates of the 1-chain $(x_a, x_b, x_c, x_d, x_e, x_f, x_g, x_h)^T$ if and only if the 1-chain is a 1-cycle of the network. Give a set of four linear equations satisfied by coordinates of the 1-cochain $(y_a, y_b, y_c, y_d, y_e, y_f, y_g, y_h)$ if and only if 1-cochain is a 1-coboundary.

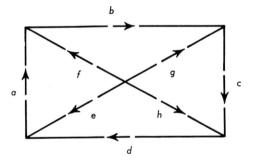

Fig. 6.11

2.

$$
\begin{bmatrix}
-1 & 0 & 0 & 0 & -1 & 0 & 0 & 1 & 0 & 0 & 0 & 0 \\
0 & -1 & 0 & 0 & 1 & -1 & 0 & 0 & 0 & 0 & 0 & 0 \\
0 & 0 & -1 & 0 & 0 & 1 & -1 & 0 & 0 & 0 & 0 & 0 \\
0 & 0 & 0 & -1 & 0 & 0 & 1 & -1 & 0 & 0 & 0 & 0 \\
1 & 0 & 0 & 0 & 0 & 0 & 0 & 0 & -1 & 0 & 0 & 1 \\
0 & 1 & 0 & 0 & 0 & 0 & 0 & 0 & 1 & -1 & 0 & 0 \\
0 & 0 & 1 & 0 & 0 & 0 & 0 & 0 & 0 & 1 & -1 & 0 \\
0 & 0 & 0 & 1 & 0 & 0 & 0 & 0 & 0 & 0 & 1 & -1
\end{bmatrix}
.
$$

Represent the network with this incidence matrix by a diagram. Select a maximal tree and use the different symbols in your diagram for the in-tree and out-of-tree edges. Write down the circuit $\omega(t)$ for each out-of-tree edge t. Give the dual sets of linear equations whose solutions are the 1-cycles and 1-coboundaries.

Section 6.5

1. Draw a diagram of the network with incidence matrix

$$\begin{array}{cccc} a & b & c & d \\ \end{array}$$
$$\begin{bmatrix} 1 & -1 & -1 & -1 \\ -1 & 1 & 1 & 1 \end{bmatrix} \begin{matrix} A \\ B \end{matrix}.$$

If $[w_a, 0, 0, 0]$ is the 1-cochain of impressed electromotive forces and

$$\begin{bmatrix} r_a & 0 & 0 & 0 \\ 0 & r_b & 0 & 0 \\ 0 & 0 & r_c & 0 \\ 0 & 0 & 0 & r_d \end{bmatrix}$$

is the resistance matrix, find the current 1-cycle and a potential 0-cochain.

2. (Figure 6.12.) The following table gives the resistance and impressed electromotive force in each edge of the network.

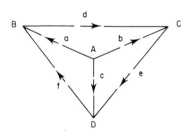

Fig. 6.12

	a	b	c	d	e	f
ohms	10	15	5	20	10	10
volts	20	−10	0	−35	45	−10

Find the current in each edge. If the potential at A is 0, find the potential at B, C, and D.

Section 6.6

1. Three warehouses, A, B, and C, contain 17, 15, and 11 units, respectively, of a commodity. Three markets, D, E, and F require 9, 12, and 22 units, respectively. The following table gives the shipping costs per unit.

From	To	Cost per Unit (dollars)	From	To	Cost per Unit (dollars)	From	To	Cost per Unit (dollars)
A	D	1	B	D	4	C	D	5
A	E	4	B	E	3	C	E	6
A	F	2	B	F	2	C	F	7

Show that the total shipping cost is minimized if no units are shipped from A to E, from B to D, from C to D, or from C to F. What is the optimal shipping pattern?

2. If the shipping cost from A to D in Exercise 1 is increased by a tariff of $2 per unit, what is the new optimal shipping pattern?

7 SOME THREE-DIMENSIONAL TOPOLOGY

7.1 Three-Dimensional Manifolds

In Chapters 1 and 2 we studied the two-dimensional topology of surfaces formed by fitting polygons together. We shall now fit polyhedral solids† together and explore three-dimensional topology. By analogy to the two-dimensional case we require the following:

1. The polyhedral solids intersect only in faces, edges, and vertices.
2. The edges and vertices of the polyhedral solids are identified only to the extent required by the face identifications.
3. At most two polyhedral solids share any particular face.
4. For any partition of the set of polyhedral solids into two subsets there is always one polygon that is a face of a solid from each subset.

A space formed in this way from a finite number of polyhedral solids is called a *three-dimensional pseudomanifold*, whereas a surface is called

---

† A *polyhedral solid* is a closed and bounded convex body in three-space whose boundary consists of a finite number of polygonal faces which meet each other along edges and vertices. It is the three-dimensional analogue of a polygon.

a *two-dimensional manifold.* A manifold is distinguished from a pseudo-manifold by the requirement that all points (except boundary points) have a neighborhood topologically equivalent to a spherical solid (a ball) in a Euclidean space of the same dimension as the manifold. (Although the concept of pseudomanifold is more general, we shall consider only those formed from polyhedral solids in accordance with rules 1 to 4.) A surface satisfies the homogeneity requirement of a manifold because each point not on a boundary curve has a neighborhood that is topologically equivalent to a disk in the Euclidean plane. We shall now identify the opposite faces of a cube to give a pseudomanifold containing two exceptional points with neighborhoods that are solids bounded by projective planes rather than spheres.

In a three-dimensional Euclidean space with Cartesian coordinates x, y, and z consider the cubical solid defined by the condition max ($|x|$, $|y|$, $|z|$) ≤ 1 (Figure 7.1). Match the face on which $x = 1$ to that on which

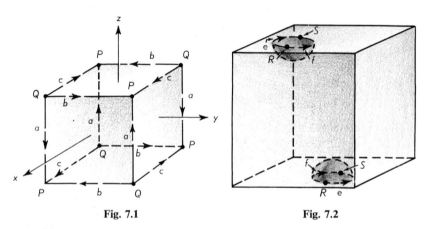

Fig. 7.1   Fig. 7.2

$x = -1$ by the correspondence $(1, y, z) \leftrightarrow (-1, -y, z)$. Similarly, identify $(x, 1, z)$ with $(x, -1, -z)$, and $(x, y, 1)$ with $(-x, y, -1)$. These face identifications impose the edge and vertex identifications indicated by the labels used in Figure 7.1. This cubical solid with identified faces is a closed pseudomanifold because the six faces are identified in pairs. Because the original solid was in Euclidean space, every point inside the cube has neighborhoods that are balls.

An interior point of a face, say the top, has spherical neighborhoods made from a hemiball from the top of the solid and a hemiball from the bottom of the solid. This is illustrated in Figure 7.2. If the bottom hemiball is reflected in a vertical plane through R and S (as they appear on the bottom face), the curves f and e are interchanged and the lower hemiball can be placed on top of the upper hemiball. Consider an interior point of an edge; for example, the point T with coordinates $(1, 1, z_0)$ on the vertical edge a (Figure 7.3). The intersection of the ball,

$$(x - 1)^2 + (y - 1)^2 + (z - z_0)^2 \leq \varepsilon,$$

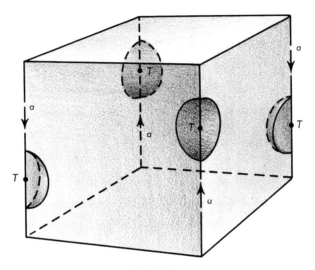

Fig. 7.3

with the cubical solid gives a spherical wedge in a neighborhood of T. An entire neighborhood of T is formed by fitting together four spherical wedges centered about the four occurrences of T. When the wedges have been fitted together, each will have one unmatched, two-sided polygonal face. The eight edges of these polygons are identified in pairs and each edge will have as endpoints the two points on a at distance ε from T. Because the Euler characteristic of the surface bounding the neighborhood of T is $\chi = 4 - 4 + 2 = 2$, the neighborhood is a solid bounded by a sphere. By letting $\varepsilon \to 0$ we see that the solid is a union of concentric spheres and is topologically a ball in Euclidean three-space. We have now found that the homogeneity condition of a manifold is satisfied at all points of the pseudomanifold, except possibly the vertices P and Q. We now show that the condition fails at these points. By symmetry it is sufficient to consider the point P.

Consider a neighborhood of P formed from four solids which are the intersections of the cubical solid and the balls of radius ε centered at the four vertices P. From Figure 7.4 we see that this neighborhood of P has a boundary surface represented by the equations

$$ghi = 1, \qquad gjk = 1, \qquad imj = 1, \qquad hkm = 1.$$

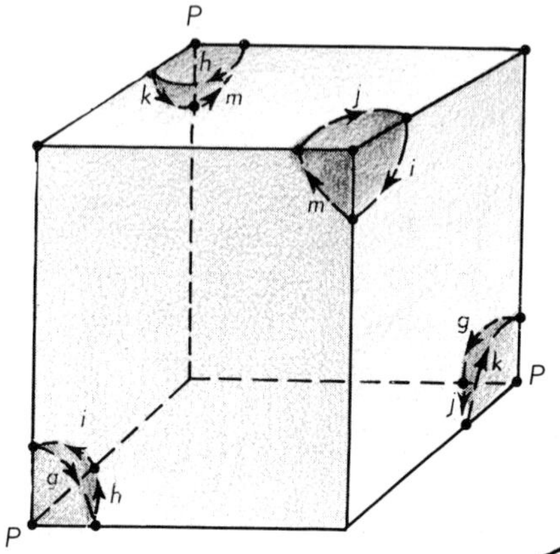

Fig. 7.4
(*Above*)

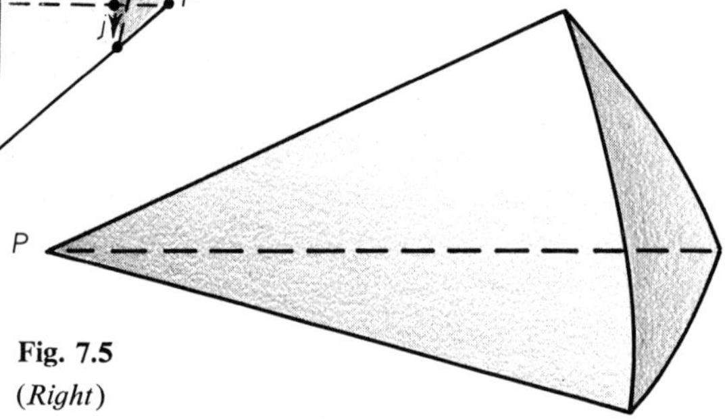

Fig. 7.5
(*Right*)

By reducing the equations to a canonical equation we find that the neighborhood of P is a solid bounded by a projective plane.

The reasoning used in this example also proves that all points, except possibly the vertices, have neighborhoods that are balls when polyhedral solids are fitted together to form a closed pseudomanifold. We shall now derive a necessary and sufficient condition for all of the vertices to have neighborhoods that are balls.

If we consider a polyhedral solid with planar faces in Euclidean three-space, we can form a neighborhood of a vertex by intersecting the solid with a ball centered at the vertex. By selecting balls that are small enough we can find a set of disjoint neighborhoods, one for each vertex. Furthermore, the neighborhood of a vertex may be assumed to be small enough to intersect only the edges and faces containing the given vertex. We call these neighborhoods *corners of the polyhedral solid.* A corner at a vertex

P is itself a polyhedral solid with one curved face. If a face of the original solid has P as a vertex, the face will contribute a circular sector as a face of the corner. The circular arcs of these sectors are the edges of a spherical polygon which is the curved face or base of the corner. There is a vertex of this base on every edge of the polyhedral solid with P as a vertex.

When polyhedral solids are fitted together to form a closed pseudo-manifold, each vertex has among its neighborhoods some that are a union of topological equivalents of polyhedral corners and are bounded by a closed surface formed from the bases of these corners. Although these statements are intuitively reasonable, an exact proof would depend on a careful definition and study of neighborhoods. Assuming that neighborhoods of this special type have been selected for each vertex, we shall show that a necessary and sufficient condition that all of these neighborhoods be bounded by spheres is that the sum of the number of polyhedral solids and the number of edges equal the sum of the number of faces and the number of vertices.

Let n_0, n_1, n_2, n_3 be the numbers of vertices, edges, faces, and polyhedral solids and n_0', n_1', n_2' are the numbers of vertices, edges, and faces if the identifications are ignored. Let m_0, m_1, m_2 be the numbers of vertices, edges, and polygons in the set of surfaces bounding the neighborhoods. Because each edge of the closed pseudomanifold meets the neighborhood surfaces at one point near each end of the edge, $m_0 = 2n_1$. The equation $n_2' = 2n_2$ follows because the faces of the polyhedral solids are identified in pairs. If identifications are ignored, the number of vertices of a polyhedral solid equals the number of corners. Hence $n_0' = m_2$. When the identifications are ignored, the base of each corner is a polygon with the same number of edges as vertices. Because edges of the polygons are identified in pairs and the edges of the polyhedral solids contain two vertices of the neighborhood surfaces, $2m_1 = 2n_1'$. The inequality

$$m_0 - m_1 + m_2 \le 2n_0$$

follows because each of the n_0 neighborhood surfaces has Euler characteristic less than or equal to 2. Because the surface of each polyhedral solid is topologically a sphere,

$$n_0' - n_1' + n_2' = 2n_3 .$$

Substitution for m_0, m_1, m_2, and n_2 gives

$$2n_1 - n_1' + n_0' \le 2n_0 ,$$
$$n_0' - n_1' + 2n_2 = 2n_3 .$$

By subtracting the equation from the inequality we derive the inequality

$$2n_1 - 2n_2 \leq 2n_0 - 2n_3,$$

which is equivalent to

$$n_3 + n_1 \leq n_0 + n_2.$$

The equality will hold if and only if

$$m_0 - m_1 + m_2 = 2n_0.$$

Because this equation holds if and only if the Euler characteristic of each neighborhood surface equals 2, we have proved the desired result.

In two-dimensional topology the Euler characteristic is useful in the classification of closed surfaces. The three-dimensional Euler characteristic defined by

$$\chi = n_0 - n_1 + n_2 - n_3$$

does not help to classify closed three-manifolds.

Theorem. A closed pseudomanifold formed from polyhedral solids in accordance with rules 1 to 4 is a manifold if and only if the Euler characteristic is zero.

7.2   Orientability

Before defining orientability of three-dimensional manifolds we shall give an example of a nonorientable manifold and show that some of its properties contradict our space intuition based on Euclidean solid geometry.

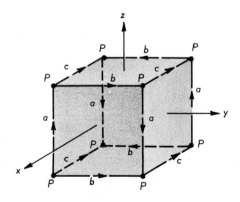

Fig. 7.6

As in an earlier example, we start with the cubical solid defined by the condition max $(|x|, |y|, |z|) \le 1$. We identify the opposite faces by the following correspondences: $(1, y, z) \leftrightarrow (-1, -y, z)$, $(x, 1, z) \leftrightarrow (x, -1, -z)$, $(x, y, 1) \leftrightarrow (x, y, -1)$. These face identifications induce the edge and vertex identifications shown in Figure 7.6. Because the Euler characteristic is $\chi = 1 - 3 + 3 - 1 = 0$, this cube with identified faces is a manifold.

Consider the midsection of this manifold on which $z = 0$. As the identifications in Figure 7.7 show, this midsection is a Klein bottle represented by the equation $dede^{-1} = 1$. In Figure 7.8 we show the Klein bottle with a

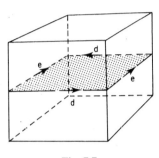

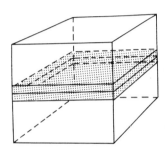

Fig. 7.7   Fig. 7.8

neighborhood, which consists of the bottle and two thin blankets, one above and one below it. In the matching of the left and right faces, the right-edge upper-blanket side is matched to the left edge of the lower blanket. Because the Klein bottle does not separate the two blankets, it is a one-sided surface. In contrast, let us investigate the Klein bottle which is the midsection on which $x = 0$ (Figure 7.9). Because the face identifications match the front half of the left face to the front half of the right face and the front half of the top face to the front half of the bottom face, the front blanket in a neighborhood of the Klein bottle in Figure 7.9 is not connected to the rear blanket. This shows that this Klein bottle is a two-sided surface. Figure 7.10 shows that the midsection in which $y = 0$ is a one-sided torus, for the left half of the front face is identified with the right half of the rear face. A torus inside the cube is two-sided. We have now seen that the one-sidedness or two-sidedness of a surface is not its intrinsic property but depends on the type of three-dimensional manifold in which it is embedded and its position in this manifold. The very definition of a nonorientable three-dimensional manifold depends on two-sided Klein bottles.

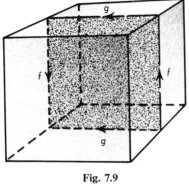
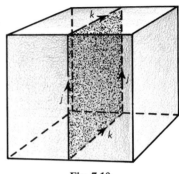

Fig. 7.9   **Fig. 7.10**

$fgfg^{-1} = 1$ $jkj^{-1}k^{-1} = 1$

Consider a neighborhood of a simple closed curve in a three-dimensional manifold. We can think of this neighborhood intuitively as a solid tube encasing the curve.† If we cut across this tube, the neighborhood becomes a cylindrical solid. Thus the original tube is a cylindrical solid with the circular disk at one end identified with the circular disk at the other. The boundary of the tube is a cylinder with the circle at one end identified with the circle at the other. This boundary surface is a torus or a Klein bottle, depending on how the orientations of the two circles are matched. The manifold is defined as *nonorientable* if it contains a closed curve with a neighborhood bounded by a Klein bottle. Note that this definition requires the existence of at least one simple closed curve with at least one neighborhood bounded by a Klein bottle. This Klein bottle is, of course, two-sided, for it separates the neighborhood from the rest of the manifold.

Because a manifold may be shown to be nonorientable by exhibiting a single two-sided Klein bottle, our definition is satisfactory for proving nonorientability. The definition in this form is poorly suited to proving orientability, for it requires showing that no tubular neighborhood, no matter how wild the curve, can be bounded by a two-sided Klein bottle. There are other equivalent definitions that are more useful. The present one was selected because it is intuitive and could be stated without additional terminology and concepts.

In the manifold of the example the intersection of the x-axis and the cubical solid is a closed curve which has a neighborhood bounded by a Klein bottle. Consider a particular tubular neighborhood of this curve.

---

† Wild curves may not have any tubular neighborhoods.

At the origin select three vectors in the tube directed positively along the coordinate axes. Slide these vectors along the x-axis so that the vectors continue to be parallel to the three axes. As the vectors disappear through the front face and reappear through the back face, two of the vectors stay parallel to the positive direction of the x- and z-axes but the third changes from the positive y-direction to the negative y-direction. With this change a triple of vectors with a right-hand orientation become a triple with a left-hand orientation. Thus it is not possible to distinguish right-hand or left-hand orientations in this nonorientable manifold.

We now describe two orientable three-dimensional manifolds. The first is the hypersphere (or three-sphere) defined in a four-dimensional Euclidean space with Cartesian coordinates x, y, z, t by the equation $x^2 + y^2 + z^2 + t^2 = 1$. In Chapter 1 we used stereographic projection to represent a plane completed by a point at infinity as a two-dimensional sphere. We reverse the procedure to represent the hypersphere as a Euclidean three-space completed by a single point at infinity. We project the point (x, y, z, t) on the hypersphere from the north pole $(0, 0, 0, 1)$ onto a point in the hyperplane (or three-space) defined by $t = 0$. To find the image of (x, y, z, t) we must find the intersection of the hyperplane $t = 0$ and the line through the points (x, y, z, t) and $(0, 0, 0, 1)$. All points on this line have the form

$$(0, 0, 0, 1) + k(x, y, z, t - 1).$$

Because the desired point of intersection has $k = 1/(1 - t)$, the image of (x, y, z, t) is $[x/(1 - t), y/(1 - t), z/(1 - t), 0]$. Computation shows that the point $(x, y, z, 0)$ in the hyperplane is the image of the point

$$\left(\frac{2x}{x^2 + y^2 + z^2 + 1}, \frac{2y}{x^2 + y^2 + z^2 + 1}, \frac{2z}{x^2 + y^2 + z^2 + 1}, \right.$$
$$\left. \frac{x^2 + y^2 + z^2 - 1}{x^2 + y^2 + z^2 + 1} \right)$$

on the hypersphere. This shows that stereographic projection from the north pole gives a one-to-one correspondence between the points of the hypersphere with the north pole deleted and the hyperplane $t = 0$. The formulas show that the correspondence is continuous in both directions. To extend the mapping to the entire hypersphere we add a point at infinity to the hyperplane as an image of the point $(0, 0, 0, 1)$. A hypersphere embedded in four-dimensional Euclidean space is topologically equivalent to Euclidean three-space completed with a single point at infinity.

Consider in the hypersphere a simple closed curve encased in a tubular neighborhood bounded by a torus or Klein bottle. By "rotating" the hypersphere we can ensure that the north pole is not in the closed tubular neighborhood. Under stereographic projection the torus or Klein bottle corresponds to a torus or Klein bottle, respectively, in Euclidean three-space. The surface must be a torus rather than a Klein bottle, for a Klein bottle cannot be embedded in Euclidean three-space. This shows that the hypersphere is orientable.

To prove that the hypersphere is a manifold we wish to divide it into polyhedral solids. Under stereographic projection the (southern) half of the hypersphere defined by $t \leq 0$ corresponds to the ball defined by $t = 0$, $x^2 + y^2 + z^2 \leq 1$. Each point on the equatorial sphere defined by $t = 0$ and $x^2 + y^2 + z^2 = 1$ corresponds to itself. Stereographic projection from the south pole is defined by the transformation $(x, y, z, t) \leftrightarrow (x/(1 + t)$, $y/(1 + t)$, $z/(1 + t), 0)$. Under this projection the half of the hypersphere with $t \geq 0$ is mapped into the same solid sphere and the points on the equatorial sphere are left fixed. The hypersphere may be considered as a two-branched covering of the ball with the branches coinciding on its spherical boundary. If the bounding sphere is divided into polygons, edges, and vertices, the hypersphere is the union of two polyhedral solids. To examine the identification of the faces of these polyhedral solids we transform their representative solid spheres so that the images of the balls no longer coincide. In discussing the balls in the hyperplane $t = 0$, the coordinate t is omitted. Use the transformation

$$(x, y, z) \rightarrow \tfrac{1}{2}(x, y, z) + \tfrac{1}{2}(x, y, \sqrt{1 - x^2 - y^2})$$

on the ball considered as a representation of the northern half of the hypersphere. The ball is compressed into the hemiball defined by $z \geq 0$, $x^2 + y^2 + z^2 \leq 1$. The (upper) hemisphere $z = \sqrt{1 - x^2 - y^2}$ is left fixed, whereas the (lower) hemisphere $z = -\sqrt{1 - x^2 - y^2}$ is projected parallel to the z-axis onto the circular disk $z = 0$, $x^2 + y^2 \leq 1$. On the ball, as a representation of the southern half of the hypersphere, perform the same transformation but follow it by the reflection $(x, y, z) \rightarrow (x, y, -z)$ in the xy-plane. Under the composite transformation the ball is mapped onto the hemiball $z \leq 0$, $x^2 + y^2 + z^2 \leq 1$. The lower hemisphere is projected as before onto the equatorial disk and the upper hemisphere is transformed into the lower hemisphere by reflection in the xy-plane. We now have a representation of the northern and southern halves of the hypersphere as the upper and lower halves of the ball $x^2 + y^2 + z^2 \leq 1$. Although a point of the disk $z = 0$, $x^2 + y^2 \leq 1$ is in both hemiballs, it represents a single point in the equatorial sphere of the hypersphere. Thus the hemiballs fit together in the same way as the halves of the hypersphere. In addition, the upper and lower hemispheres must be identified by the correspondence $(x, y, \sqrt{1 - x^2 - y^2}) \leftrightarrow (x, y, -\sqrt{1 - x^2 - y^2})$. We have found that a hypersphere may be represented by the polyhedral solid $x^2 + y^2 + z^2 \leq 1$ with two hemispherical faces: $z = \sqrt{1 - x^2 - y^2}$ and $z = -\sqrt{1 - x^2 - y^2}$, one edge: $z = 0$, $x^2 + y^2 = 1$, and one vertex: the point $(1, 0, 0)$. The faces are matched by identifying each point with its image under reflection in the xy-plane. The Euler characteristic of this polyhedral solid with identified faces is $\chi = 1 - 1 + 1 - 1 = 0$. We now know that the hypersphere is a three-dimensional, closed, orientable manifold.

The face identification that makes a solid sphere into a hypersphere is a three-dimensional analogue of an edge identification which transforms a circular disk into a sphere. Consider the circular disk $x^2 + y^2 \leq 1$ and identify the point $(x, \sqrt{1 - x^2})$ on the upper semicircle with the point $(x, -\sqrt{1 - x^2})$ on the lower semicircle of the boundary of the disk. This identification is indicated in Figure 7.11. The correspondence $(x, \sqrt{1 - x^2}) \leftrightarrow (-x, -\sqrt{1 - x^2})$, which matches antipodal points on the circumference of the disk, changes the disk into the projective plane (Figure 7.12). In

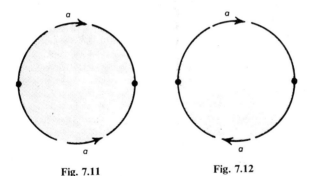

Fig. 7.11   Fig. 7.12

three dimensions the identification of antipodal points on the sphere $x^2 + y^2 + z^2 = 1$ transforms the ball $x^2 + y^2 + z^2 \leq 1$ into a three-dimensional closed manifold topologically equivalent to projective three-space.

The projective plane is formed from the Euclidean plane by adding a point at infinity on each family of parallel lines. The point at infinity on a line is reached by traveling along the line in either direction. This means a projective line is topologically equivalent to a circle. The points at infinity form the line at infinity. Three-dimensional projective space is constructed from Euclidean three-space by adding a plane at infinity that contains one point at infinity for each family of parallel lines in the Euclidean space. The transformation

$$(x, y, z) \rightarrow \left(\frac{x}{1 + r}, \frac{y}{1 + r}, \frac{z}{1 + r} \right),$$

where $r = \sqrt{x^2 + y^2 + z^2}$, maps the Euclidean space with Cartesian coordinates x, y, z in a one-to-one fashion onto the open ball defined by $x^2 + y^2 + z^2 < 1$. The radii of this ball are the images of the lines through the origin. The points on the sphere $x^2 + y^2 + z^2 = 1$ at the ends of the radii are a concrete representation of the points at infinity on the lines through the origin in the projective extension of the Euclidean space. These are all the points at infinity, for every family of parallel lines contains one line through the origin. Because the same point at infinity is reached by traveling in either direction on a line, the antipodal points of the sphere $x^2 + y^2 + z^2 = 1$ must be identified. Thus projective three-space is topologically equivalent to a ball with pairs of antipodal boundary points identified.

We now ask whether projective three-space is orientable or nonorientable.

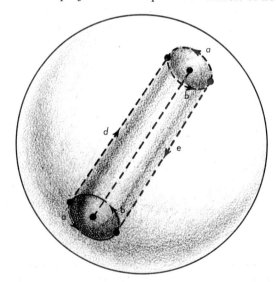

Fig. 7.13

If a closed tubular neighborhood of a simple closed curve stays entirely inside the ball, the neighborhood is bounded by a torus. Of more interest is a tubular neighborhood, which reaches the surface of the ball and then reappears at the opposite side of the surface. The example sketched in Figure 7.13 should convince us intuitively that projective three-space is orientable.† The surface bounding the tubular neighborhood of the curve c may be represented by the equations $bda^{-1}e = 1$ and $ae^{-1}b^{-1}d^{-1} = 1$. Because these equations show that the surface is orientable, the neighborhood is bounded by a torus rather than a Klein bottle.

Our examples have all been closed pseudomanifolds or manifolds. We could remove portions from a closed manifold to create boundary surfaces as we removed circular patches to give boundary curves on a surface. We might then try to generalize the classification of surfaces. No one has yet succeeded in classifying three-manifolds.

7.3   Manifolds of Configurations

The definition of manifolds makes them locally Euclidean in that each point has neighborhoods that are topologically equivalent to a Euclidean ball (disk in two dimensions). We now consider geometric realizations of a manifold such that the " points " of the manifold are lines, circles, triangles, or other configurations in Euclidean space. To fit configurations together to form a manifold we shall define the distance between configurations in each example. This permits us to discuss convergence for sequences of configurations. We shall represent a manifold of configurations as a manifold of ordinary points by giving a mapping of the set of configurations onto a set of points so that a sequence of configurations will converge if and only if the corresponding sequence of representing points converges.

As our first example, consider the diameters of a sphere. Define the distance between two diameters as the angle between them. The mapping which transforms each diameter into its pair of endpoints represents the manifold of diameters as the sphere with pairs of antipodal points identified. As an alternative we could map the diameter into its endpoint in the northern hemisphere. This would show the manifold to be topologically equivalent to a hemisphere with the antipodal points on its equator identified. With this identification the hemisphere could be represented by

---

† A rigorous proof based on our definition would be very complicated. One difficulty is that a tubular neighborhood could jump from the front to the back of the ball an infinite number of times.

the equation $aa = 1$. The manifold of diameters of a sphere is a projective plane.

As a second example, consider the tangents and secants of the circle with equation $x^2 + y^2 = 1$. Represent these secants and tangents by the chords and degenerate point chords in which these lines intersect the circular disk $x^2 + y^2 \leq 1$. Define the distance between two chords as the sum of the distance between their midpoints and the angle between the corresponding secants or tangents. This distance between midpoints taken by itself is not a satisfactory distance between chords because any two diameters have the same midpoint.

The midpoint P of a chord is on a radius OA perpendicular to the chord (Figure 7.14). If we tried to represent the chord by P, we would find that

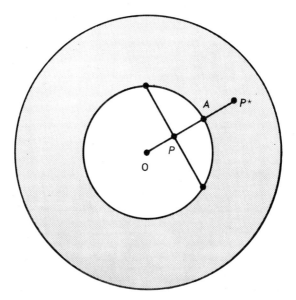

Fig. 7.14

all diameters were represented by O. To avoid this difficulty, we associate with the chord the point P^* such that A is the midpoint of P and P^*. For a degenerate chord the points P, A, and P^* all coincide. If the chord is a diameter, either half of the perpendicular diameter can serve as OA. Thus a pair of antipodal points on the circle $x^2 + y^2 = 4$ corresponds to a single diameter. The association of chords and points represents the manifold of chords as an annulus with pairs of antipodal points on the outer boundary identified. This is a projective plane with a disk removed. Thus the manifold of chords of a circle is a Möbius band.

In the example just studied the chords were unoriented, and the angles between them were between 0 and $\pi/2$. The oriented chords of a circle form a different manifold. To each proper chord there correspond two oriented chords, but to each degenerate chord there corresponds only one oriented chord, for a point does not have two distinguishable orientations. The angle between two proper oriented chords may be restricted to the interval from 0 to π. Suppose P_1 and P_2 are the midpoints of two oriented chords. Let r_1 and r_2 be the distances of P_1 and P_2 from the center of the circle. If d is the distance between P_1 and P_2 and θ is the angle between the oriented chords, define the distance between the oriented chords as $d + (1 - r_1)(1 - r_2)\theta$. Because $(1 - r_1)(1 - r_2) = 0$ if either chord is degenerate, there is no need to define the angle between oriented chords when one is degenerate.

We now describe a mapping of the oriented chords onto the annulus defined by $\frac{1}{4} \leq x^2 + y^2 \leq 4$. Let OA be the radius perpendicular to the chord at a point P. If the oriented chord crosses OA from right to left as viewed from O along OA, let the image of the oriented chord be the point P^* such that A is the midpoint of P and P^*. If the oriented chord crosses from left to right, let P^* be the midpoint of P and A. Either half of the diameter perpendicular to an oriented diameter can serve as OA. Because the oriented diameter crosses one of these radii from right to left and the other from left to right, an oriented diameter corresponds to two antipodal points, one on the outer boundary and one on the inner boundary of the annulus $\frac{1}{4} \leq x^2 + y^2 \leq 4$. This mapping represents the oriented chords as an annulus with identified pairs of antipodal points, one from each boundary. Figure 7.15 shows that this annulus may be represented by the equation $aba^{-1}b^{-1} = 1$. We have found that the manifold of oriented chords of a circle is a torus.

The three-dimensional analogue of a chord of a circle is a circular disk inscribed in a sphere. A disk inscribed in the sphere $x^2 + y^2 + z^2 = 1$ is the intersection of a plane and the ball $x^2 + y^2 + z^2 \leq 1$, whereas a chord of the circle $x^2 + y^2 = 1$ is the intersection of a line and the disk $x^2 + y^2 \leq 1$. The inscribed disk is represented by its circular boundary. The circles on a sphere should form the three-dimensional analogue of the manifold of chords of a circle. A point on the sphere is a degenerate circle obtained by intersecting a sphere and a tangent plane. Define the distance between two circles on a sphere as the sum of the distance between their centers and the angle between their planes.

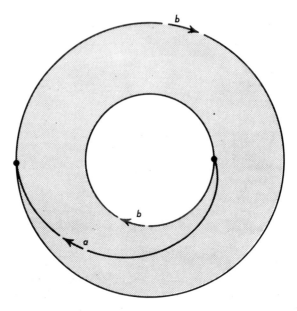

Fig. 7.15

Let P be the center of a circle on a sphere, and let OA be the radius of the sphere perpendicular to the plane of the circle. Associate with the circle the point P^* such that A is the midpoint of P and P^*. If the circle is degenerate, P, A, and P^* coincide. Because either half of the diameter perpendicular to the plane of a great circle can serve as OA, each great circle corresponds to a pair of antipodal points of the sphere $x^2 + y^2 + z^2 = 4$. This correspondence represents the manifold of circles on a sphere as the spherical shell $1 \leq x^2 + y^2 + z^2 \leq 4$ with identified pairs of antipodal points on the outer boundary sphere. This is topologically equivalent to a projective three-space with a solid sphere removed. Intuition suggests this result, for the manifold of chords of a circle is a projective plane with a disk removed. Intuition is not so helpful in identifying the manifold of oriented circles on a sphere.

There are two oriented circles for each nondegenerate circle on a sphere. We shall orient the plane of an oriented circle by defining a positive direction for the normals to the plane. A direction of a normal is positive if an observer looking in that direction reports that the circle has clockwise orientation. The angle between two oriented planes is the angle between the positive directions of their normals. Suppose that the centers P_1 and P_2

of two oriented circles on the sphere $x^2 + y^2 + z^2 = 1$ are at distances r_1 and r_2 from the center of the sphere. If d is the distance between P_1 and P_2 and θ is the angle between the oriented planes of the oriented circles, define the distance between the oriented circles as $d + (1 - r_1)(1 - r_2)\theta$. Only one oriented circle corresponds to a degenerate point circle, and this point circle cannot be used to define a positive direction normal to the corresponding tangent plane of the sphere. Therefore θ has two possible values if one of the oriented circles is degenerate. This causes no problem, however, for $(1 - r_1)(1 - r_2) = 0$ if either circle is degenerate.

We now map the oriented circles on the sphere onto the spherical shell defined by $\frac{1}{4} \le x^2 + y^2 + z^2 \le 4$. Let OA be the radius of the sphere perpendicular to the oriented plane of an oriented circle with center P. If the direction from O to A is positive with respect to the oriented plane, map the oriented circle onto the point P^* such that A is the midpoint of P and P^*. If the direction from A to O is positive, the image of the oriented circle is the midpoint P^* of P and A. Two possible radii OA correspond to an oriented great circle. Because the two radii are oppositely directed, the two image points of an oriented great circle are antipodal points, one on the outer boundary and one on the inner boundary of the spherical shell $\frac{1}{4} \le x^2 + y^2 + z^2 \le 4$. The manifold of oriented circles on a sphere is topologically equivalent to a spherical shell with identified pairs of antipodal points, one from each boundary sphere.

Because the torus of oriented chords is analogous to the manifold of oriented circles on a sphere, we might be tempted to call our new manifold a three-dimensional torus. In Section 7.4 we describe an orientable manifold that is more deserving of this name. We shall now see that the manifold of oriented circles on a sphere is nonorientable.

Figure 7.16 shows a simple curve joining $(0, 0, 2)$ and $(0, 0, -\frac{1}{2})$ in the spherical shell $\frac{1}{4} \le x^2 + y^2 + z^2 \le 4$. This curve may be considered as a simple closed curve in the manifold of oriented circles on a sphere if we identify points by the correspondence

$$\left(x, y, \sqrt{4 - x^2 - y^2}\right) \leftrightarrow \left(\frac{-x}{4}, \frac{-y}{4}, \frac{-1}{4}\sqrt{4 - x^2 - y^2}\right).$$

A tubular neighborhood of this curve is sketched in Figure 7.17. Because the boundary of this tubular neighborhood is a Klein bottle with the equation $abab^{-1} = 1$, the manifold of oriented circles on a sphere is nonorientable.

Another representation of this manifold is a hyperquadric in four-dimensional projective geometry. Suppose x_1, x_2, x_3, x_4 are Cartesian

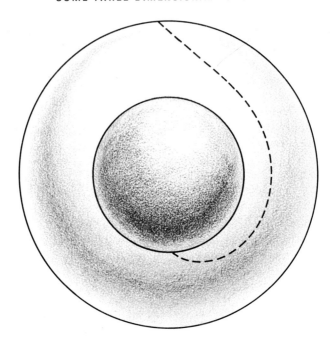

Fig. 7.16

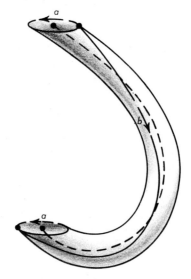

Fig. 7.17

coordinates in a four-dimensional Euclidean space. Replace these co-ordinates by the *homogeneous coordinates* t_1, t_2, t_3, t_4, t_5, where $x_1 = t_1/t_5$, $x_2 = t_2/t_5$, $x_3 = t_3/t_5$, and $x_4 = t_4/t_5$. These coordinates are said to be homogeneous because the coordinate quintuple $(ct_1, ct_2, ct_3, ct_4, ct_5)$, where $c \neq 0$ represents the same point as $(t_1, t_2, t_3, t_4, t_5)$. The quintuples with $t_5 = 0$ do not represent points in the Euclidean four-space. A line L through the point (a_1, a_2, a_3, a_4) with direction vector (b_1, b_2, b_3, b_4) has the parametric equation

$$(x_1, x_2, x_3, x_4) = (a_1, a_2, a_3, a_4) + n(b_1, b_2, b_3, b_4).$$

In terms of the homogeneous coordinates, this equation becomes

$$(t_1, t_2, t_3, t_4) = t_5(a_1, a_2, a_3, a_4) + s(b_1, b_2, b_3, b_4)$$

and is satisfied by the quintuples of homogeneous coordinates of points on L. In addition, the equation is satisfied by the quintuples

$$(t_1, t_2, t_3, t_4, t_5) = s(b_1, b_2, b_3, b_4, 0),$$

which are defined as coordinate quintuples for a point at infinity on L. Notice that this point at infinity depends only on the direction of L. Hence all lines parallel to L pass through the same point at infinity. After the introduction of these points at infinity, every quintuple of numbers except $(0, 0, 0, 0, 0)$ is a set of homogeneous coordinates for some point. The Euclidean four-space with one point at infinity added for each family of parallel lines is four-dimensional projective space.

We shall now show that the manifold of oriented circles on a sphere is topologically equivalent to the locus of the equation

$$t_1^2 + t_2^2 + t_3^2 = t_4^2 + t_5^2$$

in projective four-space. Divide the locus into four parts S_1, S_2, S_3, S_4, so that $|t_4| \leq |t_5|$ and $t_4 t_5 \geq 0$ on S_1, $|t_4| \geq |t_5|$ and $t_4 t_5 \geq 0$ on S_2, $|t_4| \leq |t_5|$ and $t_4 t_5 \leq 0$ on S_3, and $|t_4| \geq |t_5|$ and $t_4 t_5 \leq 0$ in S_4. Because the solution $t_1 = t_2 = t_3 = t_4 = t_5 = 0$ does not correspond to a point, we may assume that $t_i \neq 0$ for some i. From the equation we see that either $t_4 \neq 0$ or $t_5 \neq 0$. Because $|t_4| \leq |t_5|$ on S_1 and S_3, $t_5 \neq 0$ throughout S_1 and S_3. Similarly, $t_4 \neq 0$ throughout S_2 and S_4.

The mapping

$$(t_1, t_2, t_3, t_4, t_5) \rightarrow (x, y, z) = \frac{1}{t_5}(t_1, t_2, t_3)$$

carries S_1 into Euclidean three-space. Because

$$\left(\frac{t_1}{t_5}\right)^2 + \left(\frac{t_2}{t_5}\right)^2 + \left(\frac{t_3}{t_5}\right)^2 = \left(\frac{t_4}{t_5}\right)^2 + 1$$

and t_4/t_5 ranges from 0 to 1, S_1 is mapped onto the spherical shell $1 \le x^2 + y^2 + z^2 \le 2$. A point in S_1 is determined by the four ratios t_1/t_5, t_2/t_5, t_3/t_5, t_4/t_5, and the equation determines the fourth ratio in terms of the other three. Thus a point in the spherical shell corresponds to a unique point in S_1, which shows that the mapping is one-to-one. Continuity of the mapping and its inverse can be proved from the equations relating x, y, z and the t_i. We have found a representation of S_1 as a spherical shell.

Similarly the transformation

$$(t_1, t_2, t_3, t_4, t_5) \rightarrow (x_1, y_1, z_1) = \frac{1}{t_5}(t_1, t_2, t_3)$$

maps S_3 onto the same spherical shell. We follow this by the inversion in the sphere $x^2 + y^2 + z^2 = 1$ defined by

$$(x_1, y_1, z_1) \rightarrow (x, y, z) = \frac{1}{r^2}(x_1, y_1, z_1)$$

where $r^2 = x_1^2 + y_1^2 + z_1^2$. The composite mapping

$$(t_1, t_2, t_3, t_4, t_5) \rightarrow (x, y, z) = \frac{t_5^2}{t_4^2 + t_5^2}\left(\frac{t_1}{t_5}, \frac{t_2}{t_5}, \frac{t_3}{t_5}\right)$$

represents S_3 as the spherical shell $\frac{1}{2} \le x^2 + y^2 + z^2 \le 1$; S_1 and S_3 have a common boundary on which $t_4 = 0$. On this boundary the transformations of S_1 and S_3 agree. The mapping

$$(t_1, t_2, t_3, t_4, t_5) \rightarrow (x_1, y_1, z_1) = \frac{1}{t_4}(t_1, t_2, t_3),$$

followed by the inversion

$$(x_1, y_1, z_1) \rightarrow (x, y, z) = \frac{2}{x_1^2 + y_1^2 + z_1^2}(x_1, y_1, z_1)$$

in the sphere $x^2 + y^2 + z^2 = 2$, gives the mapping

$$(t_1, t_2, t_3, t_4, t_5) \rightarrow (x, y, z) = \frac{2t_4^2}{t_4^2 + t_5^2}\left(\frac{t_1}{t_4}, \frac{t_2}{t_4}, \frac{t_3}{t_4}\right)$$

of S_2 onto the spherical shell $2 \le x^2 + y^2 + z^2 \le 4$. S_1 and S_2 have a boundary on which $t_4 = t_5$. On this boundary the mappings of S_1 and S_2

agree. Transform S_4 onto the spherical shell $\frac{1}{2} \le x^2 + y^2 + z^2 \le 1$ by the mapping

$$(t_1, t_2, t_3, t_4, t_5) \rightarrow (x_1, y_1, z_1) = \frac{t_4^2}{t_4^2 + t_5^2} \left(\frac{t_1}{t_4}, \frac{t_2}{t_4}, \frac{t_3}{t_4} \right).$$

By inverting in the sphere $x^2 + y^2 + z^2 = \frac{1}{2}$ transform this shell into the shell $\frac{1}{4} \le x^2 + y^2 + z^2 \le \frac{1}{2}$. Finally, reflect in the origin by transforming (x, y, z) into $(-x, -y, -z)$. The product of these mappings is defined by

$$(t_1, t_2, t_3, t_4, t_5) \rightarrow (x, y, z) = -\frac{1}{2} \left(\frac{t_4^2}{t_4^2 + t_5^2} \right)^2 \left(\frac{t_1}{t_4}, \frac{t_2}{t_4}, \frac{t_3}{t_4} \right).$$

On the common boundary of S_3 and S_4, where $t_4 = -t_5$, the mappings of S_3 and S_4 agree.

The four regions S_1, S_2, S_3, S_4 are mapped onto four concentric spherical shells. The regions S_2 and S_4, which correspond to the outer and inner shells, have a common boundary, where $t_5 = 0$. This boundary is transformed onto the sphere $x^2 + y^2 + z^2 + 4$ by the mapping of S_2 and onto the sphere $x^2 + y^2 + z^2 = \frac{1}{4}$ by the mapping of S_4. Because the image of $(t_1, t_2, t_3, t_4, 0)$ as a point of S_2 is a negative multiple of the image of the same point as it appears in S_4, the two images are antipodal points, one on each of the boundary surfaces of the shell $\frac{1}{4} \le x^2 + y^2 + z^2 \le 4$. We conclude that the projective locus of the equation

$$t_1^2 + t_2^2 + t_3^2 = t_4^2 + t_5^2$$

is topologically equivalent to the manifold of oriented circles on a sphere.

As another example, consider the rotations of a sphere. A rotation of a sphere is uniquely determined by the axis of the rotation and the angle of rotation about that axis. The axis cuts the sphere in a pair of antipodal points called the poles of the rotation. The angle of rotation will have two values corresponding to the two orientations of the axis. Define the angle of rotation to be between $-\pi$ and π, with the positive angles corresponding to rotations that are counterclockwise, viewed along the positive direction of the axis. This angle will be replaced by its negative if the orientation of the axis is reversed. Consider two rotations. The axes of these rotations may be oriented so that the angle between the oriented axes is between 0 and $\pi/2$. Let θ_1 and θ_2 be the angles of rotation induced by these orientations. The difference of the angles of rotation is the minimum of $|\theta_1 - \theta_2|$ and $2\pi - |\theta_1 - \theta_2|$. We define the distance between rotations as the sum of the angles between their axes and the difference between their angles of rotation.

We now associate a point P with each rotation. If the axis of a rotation with poles A and A' is oriented from A' to A, define P as the point on the line $A'A$ such that the directed distance from the center of the sphere to P is the tangent of one half the angle of rotation. If the opposite orientation is assigned to the axis, the angle of rotation and the directed distance are multiplied by -1 but the point P does not move. The association is a one-to-one correspondence of the rotations with angle less than π with the points of Euclidean three-space. The center of the sphere, which is on all possible axes of rotation, corresponds to the identity rotation. As the angle of rotation about a fixed axis approaches π, point P moves away from the origin. A point at infinity should be added on the axis to correspond to rotation with angle π. Because the same rotation is the limit as the angle approaches $-\pi$, the same point at infinity is reached by traveling in either direction along the axis. We have found that the manifold of rotations of a sphere is topologically equivalent to three-dimensional projective space.

Another representation of the manifold of rotations is by oriented tangent lines to the sphere. Consider the sphere $x^2 + y^2 + z^2 = 1$ and an oriented line L tangent to the sphere at the south pole $(0, 0, -1)$ and parallel to the positive direction of the x-axis. A rotation will carry the point of tangency into any point A. A second rotation with A as a pole will turn the tangent line in any desired direction in the plane tangent to the sphere at A. Because the rotation is uniquely determined by the image of L, there is a one-to-one correspondence between the rotations of the sphere

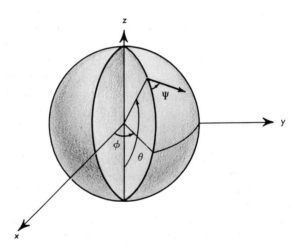

Fig. 7.18

and the oriented tangent lines to the sphere. An oriented tangent line can be described by three angles: the colatitude θ (latitude $+ \pi/2$) and longitude ϕ of the point of tangency and direction angle ψ of the oriented tangent measured counterclockwise from south to the direction of the tangent. Euler used the angles θ, ϕ, and ψ as coordinates for the rotations of the sphere. His method of defining the angles is different.

We conclude this section with a brief description of the space of closed satellite orbits in a plane. Newton considered the motion of bodies in a central force field with the magnitude of the force inversely proportional to the square of the distance from the center. He showed that a body would travel in an elliptical orbit with the center of the force field at one focus. The point of the orbit nearest to the center (perigee) and the point farthest from the center (apogee) are the vertices of the ellipse. If the two foci coincide, the ellipse is a circle and any point on the orbit can serve as perigee or apogee. The family of orbits also includes certain degenerate ellipses. These orbits are straight-line segments, with the center as perigee and apogee at the other endpoint. Of course, an earth satellite could complete only the portion of the orbit that consists of a trip directly away from the earth and a fall straight back to what would have been its starting point except for the rotation of the earth.

We let the origin be the center of the force field in a plane with polar coordinates (r, θ). We represent an orbit by a point with cylindrical coordinates (r, θ, z) in three-space, in which (r, θ) is apogee and $z = vr$ is the product of the speed at apogee v and the distance r between apogee and the center. This speed will be positive if the body is traveling counterclockwise about the origin and negative if the body is traveling clockwise. If the speed is zero the body is on the degenerate orbit of straight line travel between the origin and apogee. As the speed v increases but r remains constant, the coordinate z attains a maximum value $z(r)$ corresponding to the counterclockwise circular orbit. The z coordinate is $-z(r)$ for the clockwise circular orbit. If we were to increase z beyond $z(r)$, we would find that (r, θ) had become perigee instead of apogee. Because any point on a circular orbit can serve as apogee, all points of the circle $C(r)$ with r constant and $z = z(r)$ must be identified. Similarly, for the circle $C'(r)$ with r constant and $z = -z(r)$. The orbits correspond to points (r, θ, z) with $-z(r) \leq z \leq z(r)$ (Figure 7.19). The function $z(r)$ is an unbounded monotone increasing function of r with $z(0) = 0$.

If the set of points is deformed so that the circles $C(r)$ and $C'(r)$ shrink (to the points $(0, 0, z(r))$) and $(0, 0, -z(r))$ on the z-axis, the space of orbits

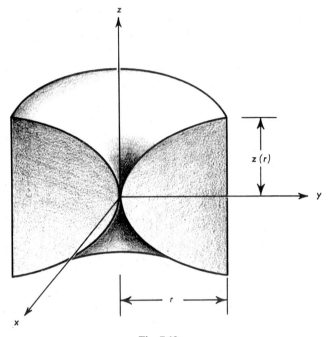

Fig. 7.19

may be represented as Euclidean three-space. By shrinking the Euclidean three-space we represent the orbits by points of the open ball $r^2 + z^2 < 1$.

Consider the points of the ball that correspond to orbits with a fixed perigee. As the speed at perigee increases, apogee moves farther away from the origin. When a certain critical speed is reached, the orbit becomes parabolic and the apogee moves to infinity. On the ball representing orbits the image of the parabolic orbit should be a point on the sphere $r^2 + z^2 = 1$. Points on the equator of this sphere correspond to the straight lines which are the degenerate parabolic orbits generated when a body moves away from the origin with "escape" velocity. Because the circular orbits have no limiting parabolic orbit, the north and south poles do not represent orbits.

We have found that the satellite orbits with the parabolic orbits included may be represented topologically by the ball $r^2 + z^2 \leq 1$, with the north and south poles deleted.

7.4 Topological Products and Fiber Bundles

With any two sets S and T, the Cartesian product $S \times T$ is the set of all ordered pairs (x, y), with $x \in S$ and $y \in T$. The plane of analytic geometry is the product of two lines; the x-axis and the y-axis. If d_S and d_T are distance functions for S and T, the distance between two pairs (x_1, y_1) and (x_2, y_2) is defined as $\sqrt{d_S(x_1, x_2)^2 + d_T(y_1, y_2)^2}$. With this definition of distance, the Cartesian product $S \times T$ is called the *topological product* of S and T. The topological product provides simple means of combining topological manifolds to form a higher dimensional manifold.

In a Euclidean three-space with Cartesian coordinates x, y, and z let I_x, I_y, and I_z be the intervals of the x-, y-, and z-axes defined by $|x| \leq 1$, $|y| \leq 1$, and $|z| \leq 1$, respectively. The topological product $I_x \times I_y$ is the rectangle in the xy-plane defined by $\max(|x|, |y|) \leq 1$. The cubical solid $\max(|x|, |y|, |z|) \leq 1$ is the topological product $(I_x \times I_y) \times I_z$. Because the topological product is associative, the parenthesis can be omitted in triple products. Let C_x, C_y, and C_z be the circles obtained by identifying the endpoints of I_x, I_y, and I_z, respectively. The product $C_x \times I_y$ is the cylinder constructed from the rectangle $I_x \times I_y$ by using the correspondence $(1, y, 0) \leftrightarrow (-1, y, 0)$ to identify one pair of opposite edges. The product $C_x \times I_y \times I_z$ is the cubical solid $I_x \times I_y \times I_z$ with the faces where $x = 1$ or $x = -1$ identified by the correspondence $(1, y, z) \leftrightarrow (-1, y, z)$. Figure 7.20 shows how the cubical solid may be deformed to achieve this identification in Euclidean space. This product is topologically equivalent to a toroidal solid. The product $C_x \times C_y$ is the torus represented by the rectangle $I_x \times I_y$ with the opposite edges identified. The product $C_x \times C_y \times I_z$ is the cubical solid $I_x \times I_y \times I_z$ with two pairs of identified opposite faces. Figure 7.21 shows how this identification can be realized in a toroidal shell.

The product $C_x \times C_y \times C_z$ is a cubical solid with the identifications $(1, y, z) \leftrightarrow (-1, y, z)$, $(x, 1, z) \leftrightarrow (x, -1, z)$, and $(x, y, 1) \leftrightarrow (x, y, -1)$. Another representation is a toroidal shell on which the pairs of points at the corresponding positions on the inner and outer boundaries are identified. Because the cubical solid with identified faces has one vertex, three edges, three faces, and one polyhedral solid, the Euler characteristic is zero. This shows that the topological product of three circles is a three-dimensional manifold. This manifold is called a *three-dimensional torus* by analogy with the usual torus, which is the product of two circles. Consider a curve in $I_x \times I_y \times I_z$ joining corresponding points on a pair of opposite faces. When the opposite faces are identified to construct $C_x \times C_y \times C_z$, the curve becomes a closed curve with a tubular neighborhood bounded

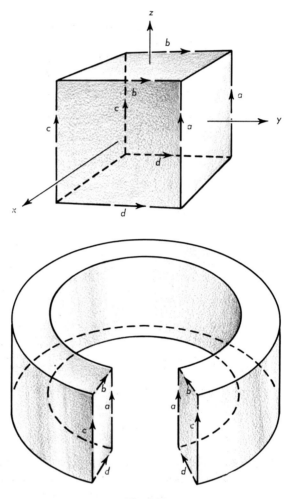

Fig. 7.20

by a torus. Intuition correctly suggests that the three-dimensional torus is orientable.

Let K_{yz} be the Klein bottle formed from $I_y \times I_z$ by the identifications $(0, 1, z) \leftrightarrow (0, -1, -z)$ and $(0, y, 1) \leftrightarrow (0, y, -1)$. The nonorientable three-manifold discussed in Section 7.1 is the topological product $I_x \times K_{yz}$.

Although the three-dimensional sphere and projective three-space are not topological products of lower dimensional manifolds, they can be

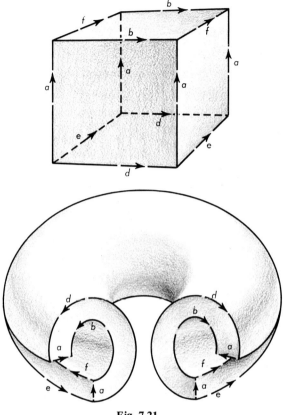

Fig. 7.21

studied by using one and two-dimensional topology. In Section 7.3 we represented projective three-space as the manifold of oriented lines tangent to the sphere $x^2 + y^2 + z^2 = 1$. The set of oriented tangents with a given point of tangency is topologically a circle. Because a circle is associated with each point of the sphere, we can describe projective three-space as the union of a sphere of circles. Consider a neighborhood N of a point P on the sphere $x^2 + y^2 + z^2 = 1$. If N is not the entire sphere, consider some point S not in N as the south pole. Every tangent to the sphere at a point in N is determined by the point of tangency, and its direction angle measured from the south. The direction angles can be considered as points on the circle formed from the real interval from $-\pi$ to π by identifying $-\pi$ and π. Thus the set of oriental lines tangent to the sphere at some point in

N is the Cartesian product of N and a circle. Our definition of the distance between rotations makes the Cartesian product equivalent to the topological product. The manifold of unit vectors tangent to the sphere is locally a topological product in the sense that the union of circles of oriented tangent lines corresponding to points in a neighborhood N of P is a topological product of N and a circle. The sphere is called the *base space*, the circle, the *fiber*, and the manifold, the *fiber bundle*. The fiber bundle is a union of disjoint fibers, each topologically equivalent to the fiber, with one fiber corresponding to each point of the base space. We have found that projective three-space is a fiber bundle with a sphere as base space and circles as fibers. It can be shown that the three-dimensional sphere is also a fiber bundle with a sphere as base space and a circle as fiber. An additional fiber bundle with the same base space and fiber would be the topological product of a sphere and a circle. Our intuitive example of a fiber bundle evaded the difficulties of giving a rigorous definition of a fiber bundle.

EXERCISES

Section 7.1

Four pseudomanifolds are formed from the cube $\max(|x|, |y|, |z|) \le 1$ by using the following correspondences to identify opposite faces:

1. $(1, y, z) \leftrightarrow (-1, y, z)$, $(x, 1, z) \leftrightarrow (x, -1, z)$, $(x, y, 1) \leftrightarrow (x, y, -1)$.
2. $(1, y, z) \leftrightarrow (-1, -y, -z)$, $(x, 1, z) \leftrightarrow (-x, -1, -z)$, $(x, y, 1) \leftrightarrow$
$$(-x, -y, -1).$$
3. $(1, y, z) \leftrightarrow (-1, z, y)$, $(x, 1, z) \leftrightarrow (z, -1, x)$, $(x, y, 1) \leftrightarrow (y, x, -1)$.
4. $(1, y, z) \leftrightarrow (-1, z, y)$, $(x, 1, z) \leftrightarrow (x, -1, -z)$, $(x, y, 1) \leftrightarrow (x, -y, -1)$.

Which of these pseudomanifolds are manifolds?

Section 7.2

1. Three manifolds are formed from the cube $\max(|x|, |y|, |z|) \le 1$ by using the following correspondences to identify opposite faces:

(a) $(1, y, z) \leftrightarrow (-1, -y, z)$, $(x, 1, z) \leftrightarrow (x, -1, z)$, $(x, y, 1) \leftrightarrow (x, y, -1)$.
(b) $(1, y, z) \leftrightarrow (-1, -y, z)$, $(x, 1, z) \leftrightarrow (-x, -1, z)$, $(x, y, 1) \leftrightarrow (x, y, -1)$.
(c) $(1, y, z) \leftrightarrow (-1, -y, -z)$, $(x, 1, z) \leftrightarrow (x, -1, z)$, $(x, y, 1) \leftrightarrow$
$$(x, y, -1).$$

Which of these manifolds are orientable?

2. The Euclidean solid S is the locus of the inequality

$$z^2 \leq (x^2 + y^2 - 1)(25 - x^2 - y^2).$$

A closed three-manifold M is formed from S by identifying a point (x, y, z) on the boundary of S with the point $(-x, -y, z)$, which is also on the boundary of S. Determine the topological nature of the closed surfaces which are the cross sections $x = 0$ and $z = 0$ of M. Are these surfaces one-sided or two-sided in M? Is M orientable?

Section 7.3

1. The manifold of lines tangent to a sphere may be constructed from the manifold of oriented lines tangent to the sphere by identifying pairs of oriented lines. Show that the manifold of tangents to a sphere is topologically equivalent to the manifold of oriented tangents to the sphere. HINT. Represent a tangent by angles θ, ϕ, ψ, where $0 \leq \psi < \pi$. Map this tangent onto the oriented tangent with Euler angles θ, ϕ, 2ψ.

2. A torus is formed by revolving the circle

$$(x - 2)^2 + y^2 = 1$$

about the y-axis. What is the topological nature of the manifold of points representing lines tangent to the torus and perpendicular to the x-axis?

3. Describe the one-dimensional pseudomanifold (network) whose points are lines (unoriented) that intersect the boundary but not the interior of a Euclidean plane triangle with three acute angles. What is the Betti number of the network? How does the Betti number change if one of the angles becomes obtuse?

Section 7.4

1. Show that the manifolds of Exercises 1a and 1b of Section 7.2 are products.

2. Let I be a line segment and C a circle.

   (a) Let S_1 be a sphere with one handle and one boundary curve and let S_2 be a sphere with three boundary curves. Show that $S_1 \times I$ is topologically equivalent to $S_2 \times I$.

(b) Let S_1 and S_2 be orientable surfaces with the same Euler characteristic and at least one boundary curve on each. Show that $S_1 \times I$
is topologically equivalent to $S_2 \times I$. Show by example that $S_1 \times C$
need not be topologically equivalent to $S_2 \times C$.

3. Let M be a Möbius band, K, a Klein bottle, C, a circle, and I, a line
segment. Prove that $M \times C$ is not topologically equivalent to $K \times I$.
HINT. Consider the boundary surfaces of these products.

BIBLIOGRAPHY

Introductory Reading

ALEKSANDROV, P. S., *Elementary Concepts of Topology*, trans. A. E. Farley. New York: Dover, 1961.

CHINN, W. G. and STEENROD, N. E., *First Concepts of Topology*, New Mathematics Library: Number 18. New York: Random House, 1966.

COURANT, R. and ROBBINS, H., *What is Mathematics?* London and New York: Oxford Univ. Press, 1948, pp 230–271.

HILBERT, D. and COHN-VOSSEN, S., *Geometry and the Imagination*, trans. P. Nemenyi. New York: Chelsea, 1952, pp 289–340.

LIETZMANN, W., *Visual Topology*, trans. M. Bruckheimer. New York: American Elsevier, 1965.

MILNOR, J. W., *Topology from the Differential Viewpoint*, notes by D. W. Weaver. Charlottesville, Va.: Univ. of Virginia Press, 1965.

STEINHAUS, H., *Mathematical Snapshots*. New York: Oxford Univ. Press, 1950, pp 214–240 or 1960, pp 272–303.

TUCKER, A. W., "Topological Properties of Disk and Sphere," *Proceedings of the First Canadian Mathematical Congress*. Toronto: Univ. of Toronto Press, 1946, pp 285–309.

More Advanced Reading

CAIRNS, S. S., *Introductory Topology*. New York: Ronald Press, 1961.

HOCKING, J. G. and YOUNG, G. S., *Topology*. Reading, Mass.: Addison-Wesley, 1961.

LEFSCHETZ, S., *Introduction to Topology*. Princeton, N.J.: Princeton Univ. Press, 1949.

WALLACE, A. H., *An Introduction to Algebraic Topology*. New York: Macmillan (Pergamon), 1957.

Motion Picture Films

(New York: Modern Learning Aids)

Differential Topology, three lectures by John Milnor (#3453, 3455, 3457).

Fixed Points, a lecture by Solomon Lefschetz (#3459).

Pits, Peaks, and Passes, two lectures by Marston Morse (#3462, 3480).

INDEX